Latrine Building

A handbook for implementing the SanPlat system

Björn Brandberg

Practical Action Publishing Ltd
25 Albert Street, Rugby, CV21 2SD, Warwickshire, UK
www.practicalactionpublishing.com

© Bjorn Brandberg 1997

First published 1997\Reprinted 2006

ISBN 13 9781853393068
ISBN Library Ebook: 9781780445281
Book DOI: http://dx.doi.org/10.3362/9781780445281

All rights reserved. No part of this publication may be reprinted or reproduced or utilized in any form or by any electronic, mechanical, or other means, now known or hereafter invented, including photocopying and recording, or in any information storage or retrieval system, without the written permission of the publishers.

A catalogue record for this book is available from the British Library.

The authors, contributors and/or editors have asserted their rights under the Copyright Designs and Patents Act 1988 to be identified as authors of their respective contributions.

Since 1974, Practical Action Publishing has published and disseminated books and information in support of international development work throughout the world. Practical Action Publishing is a trading name of Practical Action Publishing Ltd (Company Reg. No. 1159018), the wholly owned publishing company of Practical Action. Practical Action Publishing trades only in support of its parent charity objectives and any profits are covenanted back to Practical Action (Charity Reg. No. 247257, Group VAT Registration No. 880 9924 76).

Sida (Swedish International Development Co-operation Agency) is a government agency assigned to the planning, execution and evaluation of bilateral development co-operation with countries in Africa, Asia, Latin America and Eastern Europe.

Illustrations by Vic Kasinhja, Lilongwe, Malawi,
Matthew Whitton, Rugby, and the author

Typeset by My Word!, Rugby

Latrine Building

Contents

Foreword	vii
1 An introduction to better latrine building	1
2 Common types of latrine	5
3 Design and construction	25
4 Common latrine building problems and solutions	35
5 SanPlat making	53
6 Moulds for SanPlat making	67
7 Slab casting with integrated SanPlats	73
8 Handcarts	77
9 How to implement a latrine building programme	79
10 Promotion and hygiene education	87
11 How to work with people	101
12 Planning, monitoring and evaluation	115
Appendix 1 Latrine design	131
Appendix 2 Casting the small SanPlat using the all-in-one plastic mould	147
Appendix 3 Forms for sanitation programme management	153
Bibliography	161

Foreword

Faecal-borne diseases are a disaster of silent violence, taking more lives and causing more suffering than any wars and natural disasters. Unlike wars, however, they do not provide the sensationalism that attracts media attention.

Diseases such as diarrhoea, cholera and dysentery take approximately four million lives (equivalent to the loss of the population of Norway) each year, of which three million are children. This estimate does not include the innumerable cases of severe diarrhoea and parasitic infections which do not cause death but result in malnutrition and inability to fight other infections, often leading to life-long physical and mental handicaps, hampering people's ability to strive for a decent life, increasing their poverty and jeopardizing developing countries' struggle for economic growth.

The solution is simple and well known, and depends critically on three factors:

- sanitation
- water supply
- hygiene education.

Considerable progress has been achieved in water supply; but the number of families not using a sanitary latrine is increasing year by year, and one of the major reasons is cost. Investments in water supply will have little or no health impacts if not accompanied by sanitation and hygiene education. The information explosion has created an increased general knowledge and appreciation of hygiene, though practice is still lagging behind.

The SanPlat system as described in this book presents not only a low-cost sanitation technology but also a balanced approach to overcoming economic, technical, cultural and psychological constraints, where the logical solution is to ensure that water, sanitation and hygiene education are given equal attention – in budgets, in agendas, and in the number of families actually served. Promising experience in Mozambique, Malawi, Angola, Uganda, Tanzania and many other developing countries indicates that the SanPlat system makes sanitation affordable and much easier to implement.

The system builds on finding practical solutions based on what people have and are able to do using local materials and traditional skills, and mixing these with innovative approaches and new materials which stimulate the people involved.

The book gives practical recommendations with the overview as well as the details, not only on how to build but also on how to set up and implement a latrine programme – including how to work with people. It gives straightforward answers to most questions, though there will always be a need for alternative solutions.

Ingvar Andersson
Head of Africa Division
Dept of Natural Resources
   and the Environment
Sida
Stockholm

Gourisankar Ghosh
Head of Water
   and Sanitation
Unicef
New York

1. An introduction to better latrine building

Whether the aim is to build one latrine or a hundred thousand of them the design may remain the same. However, building a hundred thousand latrines involves people and planning, and makes the task a greater challenge.

This chapter provides a short introduction to the main theme of this book — building better latrines with people who want them.

What is a SanPlat?

A SanPlat is a sanitation platform which is an improved latrine slab (but not necessarily a structural slab).

SanPlats are designed with smooth surfaces to make simple latrines hygienic. They are made in fine concrete, a material which is long lasting and easy to keep clean. The concrete is formed using moulds which give correct dimensions and distinct shapes. Both the smooth concrete surfaces and the distinctive shape appeal to the eye. SanPlats have been found to be an effective incentive in latrine building campaigns.

A SanPlat latrine: hygienic and prestigious

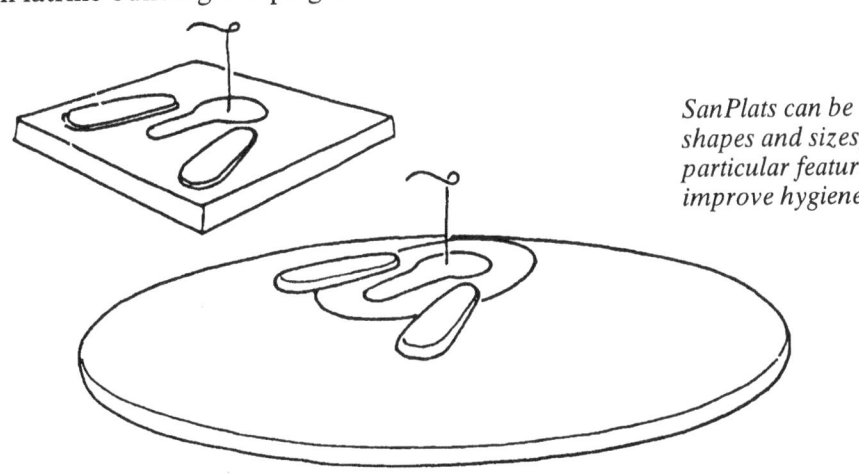

SanPlats can be of different shapes and sizes, but all have particular features which improve hygiene and safety

They have the following principal features which improve hygiene and safety:

Smooth and correctly inclined surfaces make cleaning easy

- *Elevated foot-rests* with a defined position and shape, which help the user to find the right position even at night.

  A well-studied design has essentially reduced the fouling of the squatting areas, especially in public and institutional latrines, as the people now know where to place their feet.

- *Drop-hole* shaped like a keyhole, which is safe even for a very small child.

  It is big enough to use comfortably and small enough to be completely safe.

- *A tight-fitting lid*; tight enough to shut out the foul smell, thereby also making the toilet room pleasantly odourless and free from flies.

  For public places and institutions, latrines should be fitted with ventpipes for smell and fly reduction and not have lids as the handle of the lid may cause hand-to-hand contamination.

- The SanPlat is easy to clean with water and a brush.

The SanPlat is relatively easy to transport and easy to install, requiring no special skill.

The SanPlat latrine is an improved traditional latrine. It costs less than a ventilated improved pit (VIP) latrine. The biggest advantage apart from cost is the simplicity and the possibilities it gives in scale.

Motivating people

In a community where understanding of disease transmission is limited, a health education programme should be combined with the latrine building programme. The fact that latrines can offer people privacy and convenience will also be an incentive. In addition, the idea of being modern is important for many people.

The importance of safety

A feeling of safety is very important when using a latrine. Fear that the latrine might collapse may make the user go somewhere else to attend to his or her pri-

People are your principal resource, contributing labour, ideas and organizing skills

vate needs. In this book, special attention has been given to safety. However, individual builders will need to define where the limit lies between what is safe and what is affordable.

The importance of hygiene

A dirty latrine is unpleasant to use, especially for users with bare feet. A dirty latrine is also likely to smell and attract flies. A well-designed latrine should, on the other hand, encourage cleanliness. Dirt should be easily seen and removed. For this reason careful attention needs to be paid to the construction of the latrine floor, especially around the drop-hole. A SanPlat can improve the hygiene of any latrine (including VIP and pour-flush latrines).

Fly control

The problem of flies varies from place to place. Keeping a lid on the SanPlat will make the latrine less attractive to flies, but this is not a viable option in public latrines. Putting *hot* ashes down the latrine from time to time will usually control any breeding flies.

Nobody likes the fly, but recent research has shown that it is not as dangerous as we thought

Making latrines popular and affordable

It is not until everyone uses latrines that the environment is freed from dangerous faecal matter. This can only be achieved if the latrines are both affordable and approved by the people. However, cost reduction should not take place at the expense of making the latrines pleasant to use. Improvements should be limited to the most important details, particularly the conditions around the drop-hole.

While some people may choose a large latrine, others can only afford a small one. Some people can afford a sheet metal roof and a ventpipe, but for others a simple traditional grass roof and a ventpipe may be just too complicated to build.

To encourage and spread the use of latrines, it is important that people build the types of latrine they can afford in the way they prefer so long as the results are both safe and hygienic.

In the long run, it is impossible to subsidize systems which are too expensive

Subsidies

Subsidies may be used to overcome financial constraints especially at the beginning of a programme. However, temporary subsidies can be disadvantageous, especially when used for promoting types of latrines which would otherwise not be affordable by the population.

Loans

Loans are another way to help people to pay for the materials and assistance they need to build latrines. Loans for latrines have the disadvantage of being complicated and expensive to administer. A better solution might be to combine loans for latrines in a housing or business credit scheme.

Technical advice

Though people often have their own opinions about what they want, they are usually prepared to receive good advice. Not all people are confident latrine builders, and it is the less confident groups who will appreciate assistance.

Dealing with people

Advice can be given in many ways. Treating people in a positive way usually works well, but things can go wrong. Chapter 11 of this book deals with how to work with people.

Integrated programmes

Given that the costs involved in the SanPlat system are low, and that the programme is easy to implement, the possibility and advantages of integrating it with water, health and other development programmes should always be considered.

2. Common types of latrine

What type of latrine should be built?
How complicated is it to build and use?
What are the problems related to the different types of latrines?

This chapter provides a brief overview of various types of latrines, including their construction, siting requirements, operation and maintenance, relative costs, and special problems, and will help the planner and builder when choosing from the principal options:

- traditional pit latrines
- SanPlat latrines
- conventional improved pit latrines
- VIP latrines
- pour-flush latrines
- compost latrines

For more in-depth information see the Bibliography.

The traditional pit latrine

This latrine type is well known to almost everybody throughout Africa and in many parts of Asia and Latin America.

Siting

Any latrine should always be sited some distance away from wells and boreholes, in accordance with national and World Health Organization (WHO) rules and regulations. Where possible, latrines should be sited on the lower side of sloping ground as groundwater usually flows away from a well.

Construction

The traditional latrine is a simple pit which is covered with logs, scrap material and soil. A hole is left in the middle for use and there is often a lid or slab of some kind to cover this when not in use. The traditional latrine is frequently covered with a roof to protect the pit, the cover slab and the user from rain.

For the privacy of the user the traditional latrine normally has walls and a door made of local materials such as bamboo, reeds or timber.

Although the door may be made of similar material to the rest of the structure, it is quite often made of sacking. In some cases privacy is provided by an L-shaped arrangement of walls.

In areas of collapsing soil, traditional pits may be lined with bricks, a wooden basket framework, or similar locally available materials to support the surrounding soil. (See also pages 35 to 38.)

Varieties

A large variety of traditional pit latrines exists, depending on the materials available.

- In many cases latrines are not roofed, but provided with a simple wall for privacy only. This has the advantage that a barefooted user can easily see the ground during daylight and avoid stepping on something unpleasant, even a snake (snakes often like to shelter in dark latrines). In many areas there is also a strong cultural taboo against using a 'house' as a latrine.
- Traditional latrines without walls or roofs are often

called 'night latrines' as they are normally used only during the night. Children, however, may use them during the day, and they may also be used as a safe place to deposit children's faeces which had previously been left elsewhere.
o In areas where it is difficult to dig, because of high water-tables or rocky ground, the traditional latrine may be raised or mounded to compensate for its shallow depth. This is also a way to protect the pit from being flooded.
o In parts of southern Africa (e.g. South Africa, Lesotho, Swaziland and Botswana) many traditional latrines are fitted with raised seats.
o A mini version of the traditional pit latrine is the cat pit or the one-day latrine, which is a small pit without any cover slab. It may simply be covered with a piece of scrap metal or filled directly after use.
o Traditional latrines can easily be improved with a SanPlat. This option is described later.

Not all latrines have roofs; many do not need them

Operation and maintenance

The traditional latrine is used by squatting over the drop-hole. Cleaning can be a problem as faecal matter sticks to the soil which cannot be washed. The latrine floor may need to be covered with fresh soil at regular intervals.

If wooden logs are used they may need to be replaced at intervals, depending on their type.

When the latrine pit is full it is normally abandoned and a new latrine built.

Cost and economy

The cost, in terms of money, of a traditional pit latrine is often zero as it can easily be built using readily available materials, and no special skills are required.

In many areas, however, there are people who specialize in digging pits for wells and latrines and/or covering them. In this case the whole latrine will be built by an expert who is paid to carry out the task.

In comparison with other types of latrine the traditional latrine is the cheapest. In areas where there is collapsing soil, the use of a pit lining or a pit collar may be a wise investment, even if this increases the construction cost. Pit lining is described in Chapter 3.

The cheapest option is the traditional latrine

Problems

Traditional latrines are difficult to keep clean as the floor is usually made of pounded mud. The resulting lack of hygiene has caused the traditional latrine to fall into disrepute.

The durability of the logs used may be a problem where resistant wood is difficult or expensive to acquire. Any wooden parts in contact with the ground may be attacked by rot or termites.

Mud floors can be difficult to keep clean

Rot or termite infestation may affect wooden logs

Summary

The traditional pit latrine:

- is the simplest and cheapest type of latrine as it can be made with readily available materials and without special skills;
- is very difficult to keep clean, and the humid soil around the drop-hole is an ideal breeding ground for parasites (hookworm);
- can be very easily improved with a SanPlat.

The SanPlat latrine

The SanPlat latrine is an improved version of the traditional latrine. 'SanPlat' is an abbreviation of 'sanitary platform' and is a slab (platform) with incorporated features for hygiene and child safety.

One bag of cement is enough for five to eight SanPlats. Given the low cost and the ease of implementation, the SanPlat system can easily be integrated in any development programme and rapidly reach nationwide coverage. Another advantage is that nobody will ask a family to demolish their old latrine but instead to improve it to cope with modern requirements for hygiene and health. In most cases, the cost for the improved latrine is limited to the cost of the SanPlat.

Any latrine can be improved with a SanPlat

The SanPlat is designed to fit easily into most types of latrine and especially the traditional pattern.

Siting

In common with any latrine, a SanPlat latrine should always be sited some distance away from wells and boreholes and, where possible, on the lower side of sloping ground.

The SanPlat latrine can be sited close to a house as the tight-fitting lid will prevent smell and fly nuisance.

Construction

The SanPlat is simply placed on top of the hole of the traditional latrine. If required, the surrounding earth floor may be given a new covering. The SanPlat should have:

- smooth, hard and slightly sloping surfaces to facilitate daily cleaning;
- a drop-hole shaped like a keyhole which is big enough for adults to use comfortably, yet small enough to be safe even for little children to use;
- elevated foot-rests, which help the user to find the right position even when it is dark;
- a closely fitting lid, tight enough to stop smell and flies (optional).

The SanPlat should preferably be slightly elevated above the surrounding floor so that it can be easily found in the dark. A raised SanPlat will also be cleaner, as the upstand round it will prevent wet soil from being swept over the surface when cleaning.

See Chapter 5 for information on how to make SanPlats.

Local conditions

HARD SOIL

In areas where the soil is hard and stable, unlined rectangular pits may be used. The short span over the pit makes it easy to cover with small logs. Note that thinner logs may need to be replaced at shorter intervals as they are more easily attacked by rot and termites. The superstructure should consequently be made wider that the length of the logs to allow replacement when necessary. If possible the SanPlat should be supported by pest-resistant logs or concrete lintels.

STABLE SOIL

In areas where the soil is reasonably stable, unlined, round pits can be recommended. Round pits are more stable than rectangular ones. The rectangular ones may, however, be easier to cover. Whether to make them rectangular or round may be dictated by tradition, which may have a decisive influence in these circumstances.

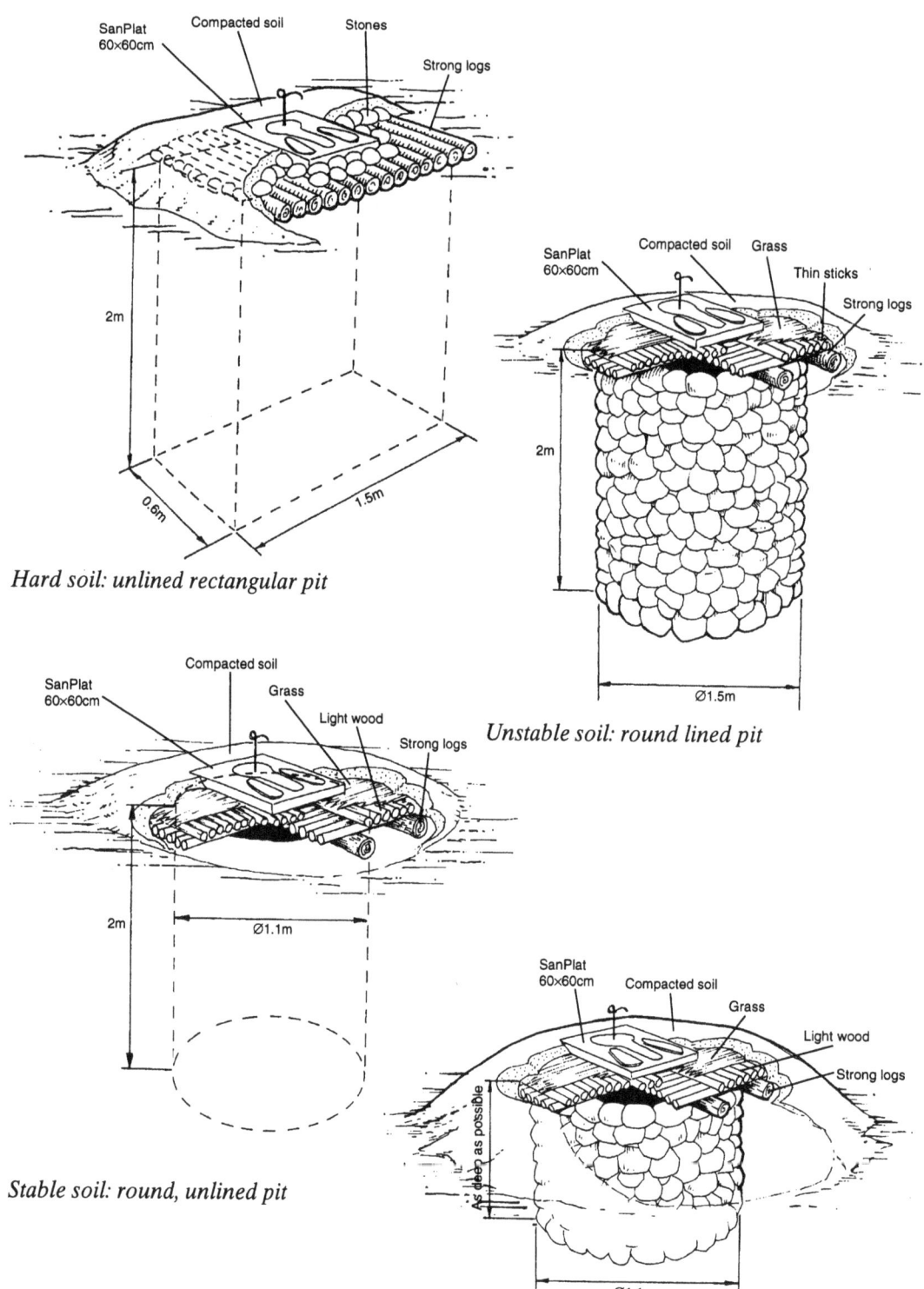

Hard soil: unlined rectangular pit

Unstable soil: round lined pit

Stable soil: round, unlined pit

Soil which is difficult to dig: raised (mounded) pit

The size of the pit should be dictated by the durability of the logs. If this is not feasible, the logs will need to be replaced from time to time.

Unstable soil

In areas where there is a risk of soil collapsing, the pit should be made round and lined. In the illustration, rocks have been used for lining the pit. If stone is not available, bricks of cement or fired bricks may be used. The top courses should be laid in full cement mortar, while the rest of the stones will be held in place by the circular shape of the pit-lining and the weight of the stones and slab.

Round pits are more stable than rectangular ones

Soil which is difficult to dig

In areas where it is difficult to dig, because of rocky soil and/or high groundwater tables, latrines can be mounted above the ground surface. In the illustration, rocks have been used for lining the pit. If stone is not available, bricks of cement or fired bricks may be used. The top courses should be laid in full cement mortar, while the rest of the stones will stay firmly in place because of the shape of the pit and the weight of the stones and slab.

Varieties

Almost any latrine can be improved with a SanPlat as a wide variety of types exists. The choice may depend on local conditions, the materials available and their cost.

A selection of the main types is illustrated here.

- Where the availability of resistant timber (or replacement logs) is a problem, the SanPlat can be made big enough to cover the whole pit. For ease of transport, the slab can be made dome-shaped. Such slabs do not normally require any reinforcement as their shape provides extra strength. A description of dome-shaped slab making is found in Chapter 5.
- Public and institutional latrines should have SanPlats without lids as their handles are liable to get dirty: in this case the latrines should be provided with ventpipes as for VIPs (see below).
- Latrines with raised seats (which are commonly found in southern Africa) can also be fitted with SanPlats even if this was not envisaged in the original design.

Operation and maintenance

SanPlat latrines are easy to use. The hard, smooth and sloping surfaces make cleaning easy as faecal deposits are easily washed down into the pit. Re-coating of the surrounding floor is no longer required as faecal matter falls only on to the SanPlat. The foot-rests help the user to find the right position, and reduce faulting to a minimum.

If wooden logs have been used in the construction they may need to be replaced at intervals depending on the type of timber used.

Cost and economy

The cost of a SanPlat varies from place to place depending on the price of materials and labour. In some places the low cost has led people to call it the 'two-dollar latrine'. In many programmes the communities provide sand and stone, while the project contributes cement, reinforcement and labour. One bag of cement is enough for up to eight SanPlats, and there is no need for a family to destroy their old latrine, so the cost of the improved latrine is only that of the SanPlat.

Affordability is a requirement for long-term sustainability

Given that a SanPlat can be moved from one latrine to another, its long-term cost is minimal. The economy of the SanPlat latrine does not therefore differ from that of the traditional latrines as the SanPlat can be reused when the old latrine is full.

Problems

It can happen that lids disappear or get broken. Replacement lids should therefore be available. Where soft logs have been used in the construction these may need to be inspected at regular intervals.

Summary

The SanPlat:

- is a slab with special features to make it easy to use and keep clean;
- can be used in any latrine;
- is easy to make with the proper moulds and training;
- is easy to transport;
- is easy to install;
- is very cheap;

o can be moved from an old latrine to a new one when the old one fills up.

Conventional improved pit latrines

A 'conventional improved pit latrine' is simply a traditional pit latrine built with permanent materials such as cement and bricks and a corrugated iron roof. This type of latrine is in common use throughout Africa, southern Asia and Latin America.

Siting
If fitted with a lid, the conventional improved latrine can be sited close to a house. Normal rules for protection of groundwater resources should be respected.

Construction
Normally a pit is dug and a concrete slab, with a hole in the middle of it, is cast on top. Logs and scrap materials are used for temporary shuttering.

A small building with walls and a roof is erected over the slab to provide protection and privacy. This building is normally made of bricks, with a sheet metal roof and a wooden door. In order to prolong the economic life of the latrine, a large-volume pit is normally recommended.

Pit linings are required where the stability of the soil is doubtful.

Varieties
The dimensions and materials used may vary according to the taste, economy and abilities of the owner, although the main aspects of the design remain the same.

The conventional improved pit latrine can be upgraded with a SanPlat, which should preferably be incorporated in the concrete slab when it is cast.

The possibility of building a VIP latrine rather than a conventional pit latrine should always be investigated.

Operation and maintenance
Conventional latrines normally present no major problems. Cleaning the floor may be difficult if the surface is rough and flat, and dirty water and urine may easily form puddles.

Problems with smells and flies can be reduced by putting hot ashes in the pit.

Latrines should not be used as a bath shelter unless

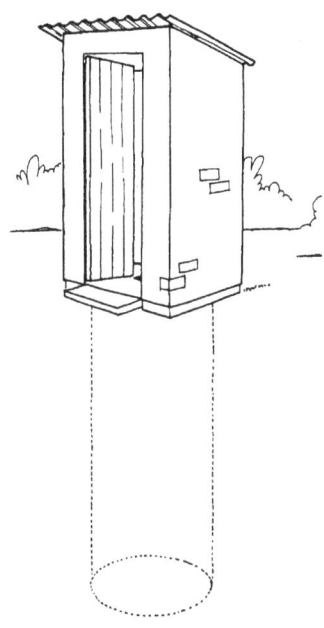

Latrines made from permanent materials need large pits to be cost-effective

separate drainage of the bath water is arranged. This is because water from the bath may either cause the soil to collapse or, if the pit is lined, cause the pit to fill quickly as the ground becomes clogged by a mixture of water, faecal matter and bacteria. Sewage water from a clogged latrine can be very dangerous.

Cost and economy

Skilled labour is required if permanent building materials like cement, bricks and iron sheets are to be used, and therefore the cash input will be considerably higher for this type of latrine than for traditional ones.

Using ashes to control smells and flies may fill the pit faster, and jeopardize its lifespan.

Problems

Normally there are few problems with conventional improved pit latrines. However, smells and flies may be a problem if there are many users. Finding space for new pits may be difficult in congested urban areas.

Summary

The conventional improved pit latrine:

o is built with permanent materials and requires skilled labour;
o is more expensive to build than a traditional latrine;
o can be difficult to replace.

The VIP Latrine

The VIP (ventilated improved pit) latrine consists of a normal pit latrine with a screened ventpipe fitted to the pit. The ventpipe should be placed straight over the pit to a height of two feet (60cm) above the roof. Flies are attracted to it by the light and are trapped in the pipe.

Siting

VIP latrines should be sited in a place where the wind will blow over the top of the ventpipe, as this is necessary for it to function properly. The latrine should be sited on the downwind side of a house as some smell will always emerge from the ventpipe.

Normal rules for protection of groundwater resources should be respected.

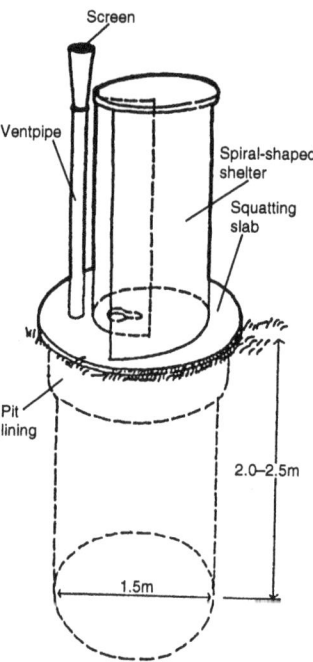

VIP latrine from Zimbabwe (Source: Winblad, 1985)

Construction

As the slab is partly unprotected by the roof, it should be cast in concrete and placed on a foundation, or a complete lining, approximately one foot (30cm) above the level of the surrounding ground to prevent rainwater from penetrating under the slab.

Start by casting the slab and allow it to cure while you make the foundation and dig the pit. Dig down until you reach stable soil and make a foundation to fit the slab.

Lay the foundation up to one foot (30cm) above the surrounding ground in brickwork and cement mortar. Allow the mortar to cure for two days before continuing to excavate the pit inside the foundation.

Use the excavated soil to fill in around the foundation and compact it well to prevent puddles forming close to the pit.

After the slab has been placed over the pit, the walls, ventpipe and roof can be erected.

Semi-darkness may be achieved by making an L-shaped entrance with or without a door.

The ventpipe must be fitted with fly-proof gauze made of stainless steel or aluminium.

Detailed recommendations for the design and construction of VIP latrines have been published by the World Bank and by Peter Morgan, of the Blair Institute in Zimbabwe (see Bibliography).

Varieties

VIP latrines can be built in many materials. Experience shows, however, that they are difficult to build with traditional materials as water from the roof may affect the mud walls and foundations. The ventpipe can be prefab-

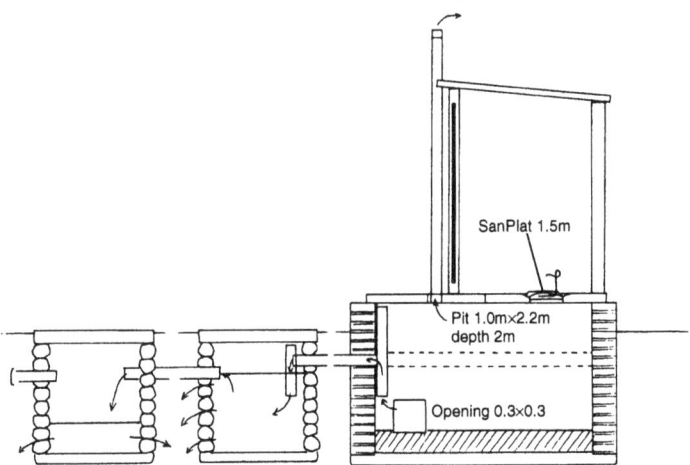

A combination of the VIP latrine and the septic tank

ricated in plastic or asbestos-cement or made with brickwork integrated with the walls. In all cases the ventpipe should be screened with a resistant fly screen.

Permanent VIP latrines can be made with a soakaway, transforming the lined pit into a septic[1] tank. In order to function properly the septic tank requires some water. This could come from the waste pipe of a bathroom.

The Blair latrine in its natural habitat (Source: Morgan, 1990). Over 1000 bricks, five or six bags of cement, and other materials are needed for the VIP latrine. Correctly built, this gives a good latrine, although it may cost more to build than some rural houses.

Drawings of septic tank based VIP latrines (SVIPs) are found in the Appendix.

Cost and economy

The normal VIP latrine uses approximately five to six bags of cement per latrine and around 1000–1200 bricks plus a gauzed ventpipe of ultraviolet-protected PVC or asbestos-cement. A low-cost version would use about three bags of cement but the same number of bricks.

If the VIP latrine is made with a door instead of an L-shaped entrance the cost may be approximately the same as for a pour-flush latrine (described in the next section).

SVIPs are more expensive but, being permanent, their long-term economy may be better. They are, however, an expensive sanitary solution.

Problems

The ventpipe, the need for a big concrete slab, and the L-shaped entrance makes the VIP latrine relatively expensive. Furthermore, the design has proven difficult to adapt to the use of traditional materials.

Foul gases from the ventpipe may require the latrine to be sited 10 metres from the dwelling house. Such

---

1 A septic tank is one in which the organic matter in sewage in disintegrated through bacterial activity. This takes up to two years, and until the process is complete the contents can cause disease.

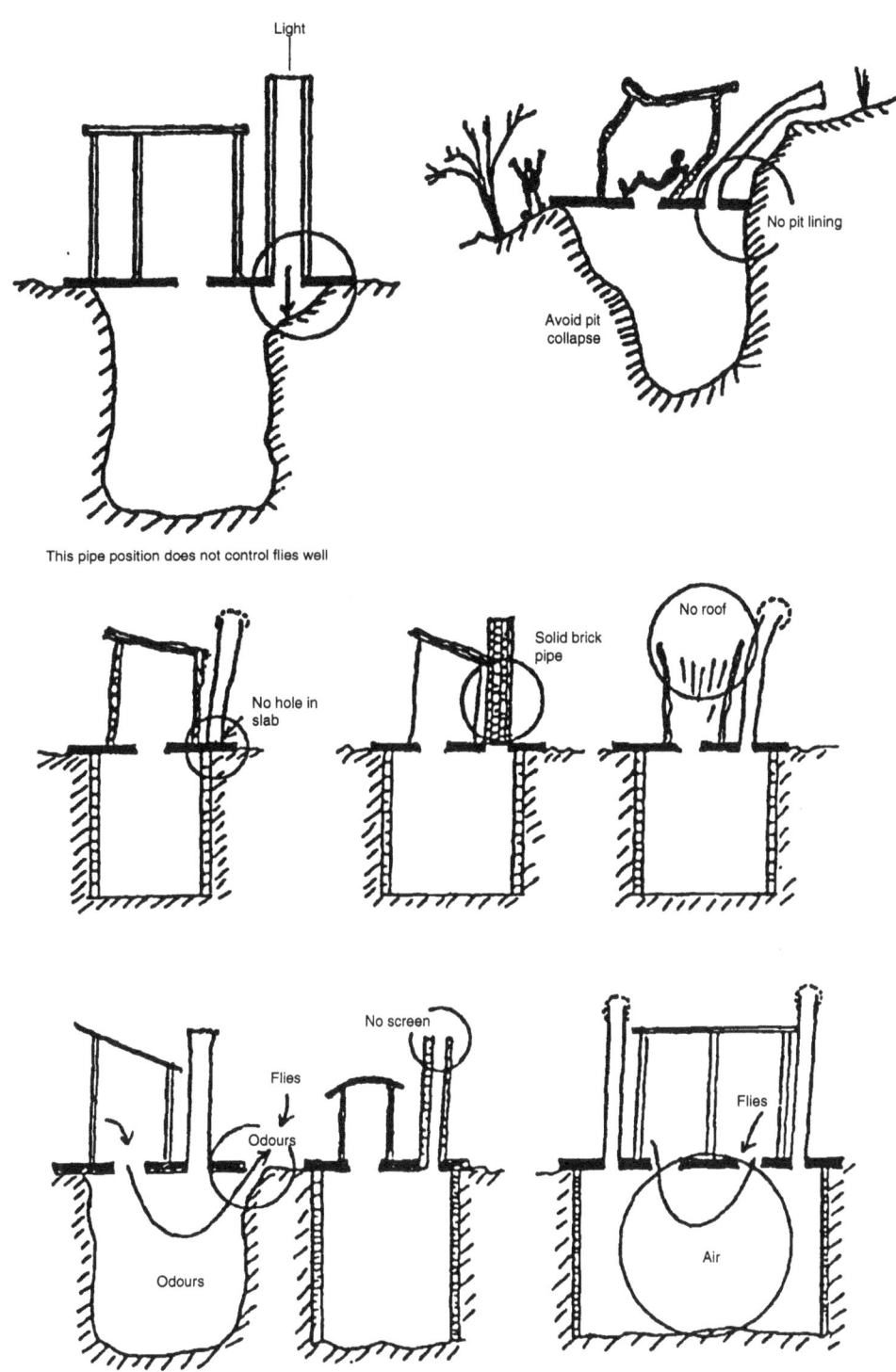

VIP latrines that cannot work (Source: Morgan, 1990) Circle denotes where error lies

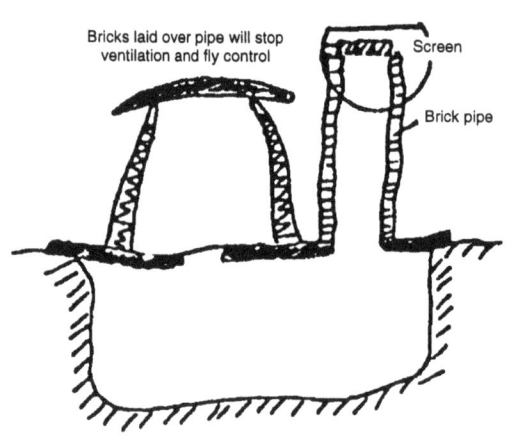

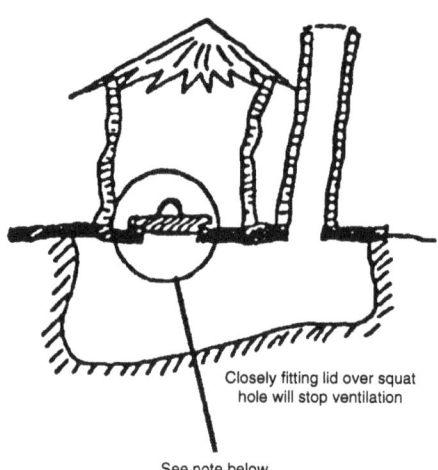

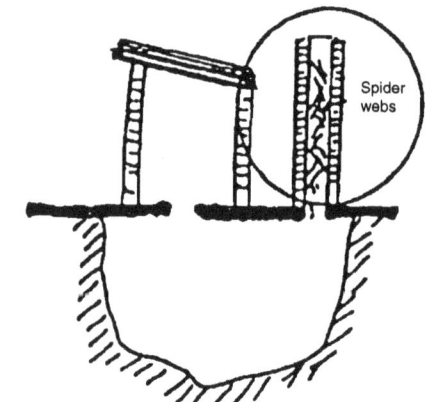

Note: *Although a closely fitting lid stops the ventilation, it also stops the smell and the flies and may well reduce these particular problems. VIP latrines with poorly fitting lids work excellently!*

VIP latrines that cannot work, continued

distances may be difficult to achieve in medium- to high-density housing areas.

Semi-darkness may discourage proper cleaning as any dirt cannot be seen easily. Dark and dirty VIP latrines do not encourage regular use because of the risk of stepping in faeces on the floor.

Resistant gauze screening material is difficult to get hold of in developing countries.

Summary
The VIP latrine:

- is free from smell and flies when correctly built;
- is difficult to adapt to traditional materials;
- is relatively expensive.

Pour-flush latrines

The pour-flush latrine is common in southern Asia. It is a traditional latrine where a water seal has been fitted to the drop-hole. It can also be built with a separate soakaway. Properly built and properly used, the water seal will prevent smell and flies coming from the latrine. In common with the VIP latrine, its cost and complexity may be a disadvantage as is the need for water for flushing the water seal.

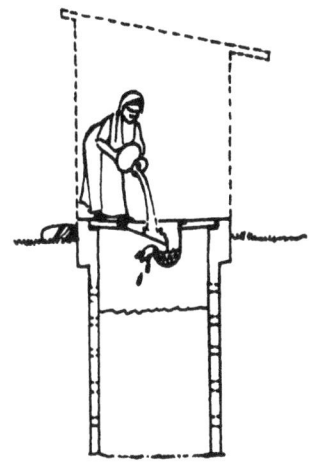

Pour-flush latrine with direct pit

Siting
A clean pour-flush latrine with a proper water seal has no smell and no flies and can be sited anywhere, even inside a house. Normal rules for protection of groundwater resources should be respected.

Construction
In most cases pour-flush latrines are built with lined pits. The lining can be made with concrete rings or brickwork. In stable soils unlined rings may be used. The slab, normally made of concrete, is placed on top of the pit.

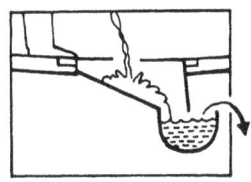

'Gooseneck' turned away from bowl

Research from Bangladesh shows that the 'gooseneck' latrine flushes better if turned away from the bowl. Reports indicate that many goosenecks have been broken but continue to function well.

The method of flushing is a serious problem as people tend to bring too little water, leaving nothing for washing hands. To get a good flush-out, the container has to be

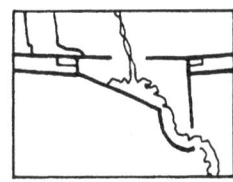

A broken gooseneck may continue to work well

tipped up rapidly, making it difficult to save some for hand washing.

Many home-made latrines use a simple polyethylene sheet hanging under the drop-hole.

In cases where the superstructure is built on top of the pit, the slab is fitted with a flushing bowl with a water seal. Very special skills are required for making this.

Varieties

The pour-flush latrine can also be made with one or two offset pits. In the case of double pits, the waste pipe from the flushing bowl has a Y-connection which can be blocked to allow alternate use of the pits, hence making it a permanent installation. Single pits may also be emptied (like septic tanks).

Operation and maintenance

If correctly designed the flushing bowl can be flushed with one or two litres of water. When the pit fills up it may be rebuilt elsewhere or emptied.

Cost and economy

The cost of a pour-flush latrine should be somewhere between that of the VIP latrine and the traditional latrine. The cost of the flushing bowl with the water seal may be about the same as the cost of a gauzed ventpipe.

Accumulation rates in pour-flush latrines should be lower than in traditional and VIP latrines and the superstructure can be smaller as an L-shaped entrance is not required.

Emptying a pour-flush latrine is probably much cheaper than rebuilding it and it is therefore a more economic design.

Problems

Blockage and breaking of the water seal is a common problem, as people often fail to use enough water. The 2-inch (5cm) water seal is only recommended where people use water for cleaning themselves, as other material will block the water seal. Emptying the latrine may be a cultural and hygienic problem.

A further constraint may be the availability of moulds and skills for making the flush pan and the water seal.

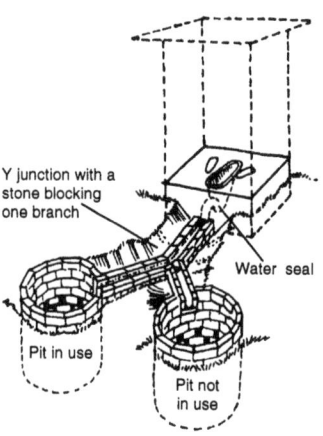

Double pit pour-flush latrine with offset pits (Source: Winblad, 1985)

Summary

The pour-flush latrine:

- o prevents smell and flies when properly built and used;
- o needs water to flush;
- o is complex and relatively expensive.

Compost latrines

General

The objective of building a compost latrine is to build a permanent latrine where the pit contents can be safely used as fertilizer. Compost latrines are normally made with two pits for alternating use, each one with an opening to allow it to be emptied. The pits should be designed

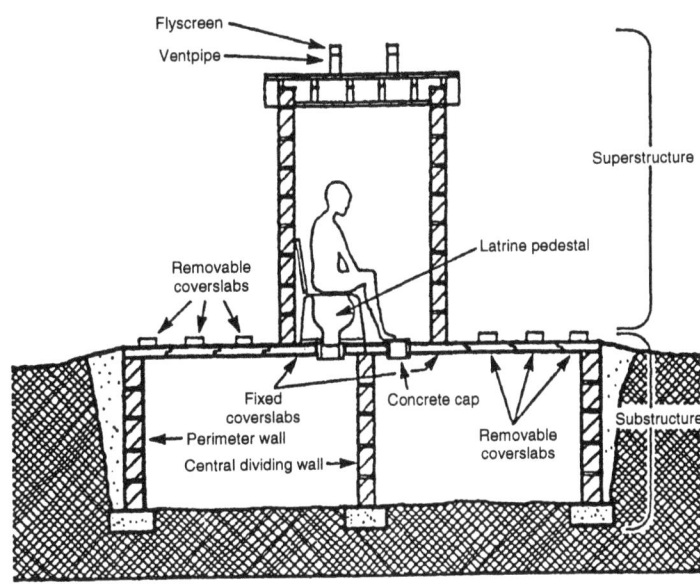

Peri-urban VIP latrine (promoted in Botswana): double pits have been built to make the latrines permanent (Source: Nostrand and Wilson, 1983)

for an accumulation time of two or more years to allow for a safe retention time and the elimination of pathogens. The estimated accumulation rate can be set to 60 litres per person per year if local data is not available.

Siting

If fitted with a ventpipe, compost latrines should be sited in a place where the wind will blow over the top of the ventpipe, as this is necessary for it to function properly. The latrine should be sited on the downwind side of a house as some smell will always emerge from the vent-pipe.

Normal rules for protection of groundwater resources should be respected.

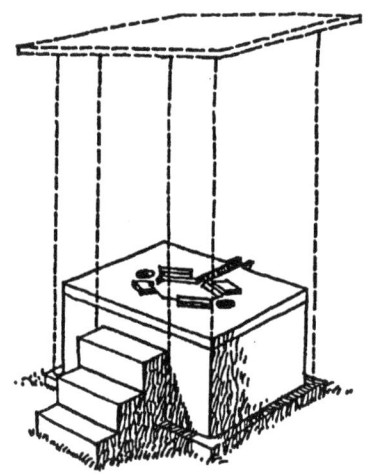

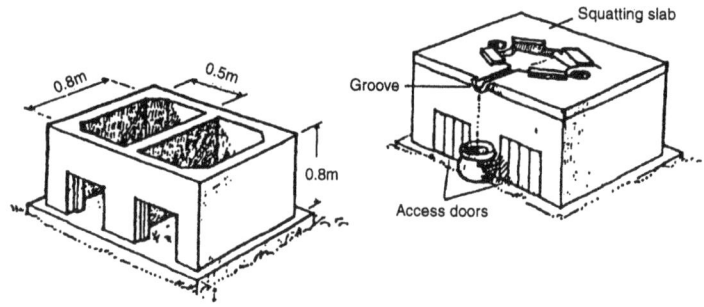

Experiences from many African countries show that it can be difficult to motivate people to empty the receptacles (Source: Winblad, 1985)

Construction

Excavation is normally shallow. The superstructure is built on top of the pits with provision being made for opening the pits.

Detailed guidelines are given in Winblad: *Sanitation Without Water*, and in the publications from World Bank and TAG Technical notes (see Bibliography).

Varieties

Most types of latrines can be built with double pits which can be emptied. Even the pour-flush latrine with a double pit is a kind of compost latrine.

Some compost latrines separate urine from faeces to prevent a slowing down of the biological decomposition.

Operation and maintenance

Before starting to use the latrine it is recommended that the pit is filled with organic material like straw or garden sweepings which will compact and decompose to a small fraction of their original volume. The compost latrine can then be used as any normal latrine. Adding more straw or garden sweepings at intervals is recommended.

When one compartment is full, the opening is sealed, and the other one taken into use. As the pits should be big enough for an accumulation time of two years, pathogens should have died off in the unused pit by the time the other one is full.

The contents of the pit can then be safely removed and used as a fertilizer, while the pit is filled with straw and taken into use again. This process can be repeated indefinitely.

Cost and economy

The cost of construction for a compost latrine is high because it has a double lined pit and often two ventpipes. Regular emptying makes it more economical as the use of the pit contents may contribute positively to the family's economy if used as a fertilizer.

Problems

- Many families, especially in Africa and southern Asia, feel reluctant to empty the pits as the contents are still considered to be faecal matter and as such untouchable.
- In many cases the pit contents are a foul wet slurry and not a soil-like compost.
- Leakage between the two pits may mean that the extracted compost contains considerable quantities of pathogens.
- The cost for the compost latrine is often prohibitive.

Summary

The compost latrine:

- requires two pits;
- is complicated and expensive to build; and
- the compost is not always safe and valued by its owner.

3. Design and construction

Even the simplest latrines can be built in different ways. Problems can easily be avoided if they are anticipated.

This chapter deals systematically with the general and technical aspects of the design and construction of latrines, considering such matters as:

- general design criteria
- planning
- the pit
- the pit lining
- the latrine slab
- the superstructure.

Drawings of the different latrines have also been included (see Appendix 1). All the examples shown include SanPlats.

General design criteria[1]

Good latrines can be made in many different ways. The best ones are normally those which do not differ too much from what people are used to building. The correct and comfortable use of the latrine is important, rather than the technology itself.

A good latrine should be:

- safe
- hygienic
- affordable
- economic
- popular!

A latrine must be liked. However well a latrine is

---
1 Design criteria: conditions which ought to be fulfilled and which you need to think of when you plan and design your latrines.

designed and built, it will never be hygienic if people cannot use it or look after it in the correct manner.

Safety

All latrines must be safe. Nobody will use a latrine if there is a risk of falling into the pit. A feeling of security is therefore a must. The risk of accidents should be kept to a harmless level.

Excessive security will, however, result in prohibitive costs. The balance between safety and affordability has to be defined by the individual families. In the end they are the ones who have to pay for the latrine and take the risk of using it. This issue should always be kept in mind.

A pit should not be too deep; it should fill up before the logs of the pit cover start to break up because of the action of rot and termites. This will eventually happen to any wood in contact with soil. Heavy walls and a thick thatched roof may cause the pit sides to collapse if the soil is weak, but a major cause of pit failure is water from heavy rains.

The risk to babies is a common reason for not having a latrine

Hygiene

All latrines should be hygienic. The principal problems of hygiene occur around the drop-hole.

The SanPlat has been designed to make the latrine hygienic. Elevated foot-rests help the user to find the right position, even in the dark. Resistant, smooth and sloping surfaces make cleaning easy. Faecal deposits, urine and water from washing should drain into the pit, while sand and floor sweepings should be swept out through the door. A tight-fitting lid and/or a ventpipe may be required for control of smell and flies. In addition, **a water container for the hygienic washing of hands is essential**.

Affordability

By promoting the use of local materials, a latrine can be affordable even for low-income households. Subsidies (if possible) to promote a latrine-building programme should only be considered for key elements such as San-Plats, ventpipes or squatting pans. Affordability is not only a question for the families. It is also a problem for the whole programme: how much can the programme afford in the future? In the final analysis, the latrine types that the community can afford will also be the latrine types the families can afford, as subsidies will not continue indefinitely.

Economy
A cheap latrine is not necessarily an economic one. Economy depends on factors such as how many years people can use the latrine. A huge pit may last for many years but may become dangerous if the pit cover is made of undersized soft timbers. In the same way a reinforced concrete slab may be very uneconomic if not reused when the pit eventually fills up. Building a permanent house over a small pit may be a bad use of both money and materials.

Popularity
The number of latrines in use is more important than their design. The design, however, is one of the most important factors influencing popularity. The intention must be to find a design which fits the culture, economy and preferences of the users. In the end it is the widespread use of a particular type of latrine which defines it as a good rather than a bad one.

Planning

Type of latrine
There are several types of latrine to choose from. Even once a type has been decided on, there are different ways of building it. The pit may be round or rectangular, and a traditional cover slab may be made of different types of logs. The roof may be sloping one, two or more ways and so on. Before starting to dig, the builder should preferably have a clear idea of what type of latrine is being built and what dimensions it should have.

Siting
Not too close to wells
Official rules for siting of latrines in relation to houses and wells should be respected wherever possible. Sometimes, however, these rules are conflicting, unrealistic or just too complicated to be practical. At least 50 feet (15m) from a well may be a good general rule.

Close to the house
A latrine far from the house will not be used. On the other hand, siting it too close may cause embarrassment.

The privacy of the entrance is especially important as many people do not like to be seen entering a latrine. Some guidelines recommend siting a latrine about 10m from the house because of smell and flies. However, with a lid there is no smell and no flies, and snakes may be a problem if the latrine is far from the house.

Downhill from a well
Wells are normally sited downhill from houses as it is easier to find water in valleys than on top of hills. Houses are frequently placed on high areas to avoid flooding, to have a better view, and to provide improved ventilation during hot weather. Placing the latrine downhill from the well, therefore, may mean that the latrine is too far away for comfortable use. A safety distance of at least 15m from the well will normally be adequate, regardless of the slope.

Downwind from the house
Wind frequently comes from different directions and it may be difficult to judge what is the most prevalent wind direction. Furthermore, what is downwind for one house may be upwind from the neighbours' houses, so a vent-pipe may be a permanent problem. Using a tight-fitting lid solves this problem.

Design
The ground is usually the best drawing board. Select the site, clear the ground and start drawing up the layout of the walls together with the plot holder. Try the design. Go in and out of the imaginary 'latrine', open and close the 'door', squat down and imagine how it would be to use the latrine you are going to build.[2] When you are ready, mark the corners with pegs as the lines will disappear when you start digging.

When you design the layout do not forget to include a place for hand washing. Decide whether a bucket of water should stand inside or outside the latrine. Where will the sullage[3] water go? Is the water protected from animals? And so on. Often it is a good idea to combine the latrine with the bath shelter. Watch, though, that the water from the bath does not drain into the latrine pit as

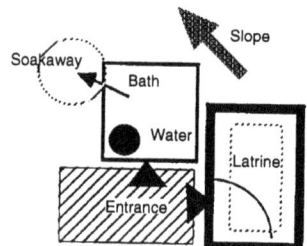

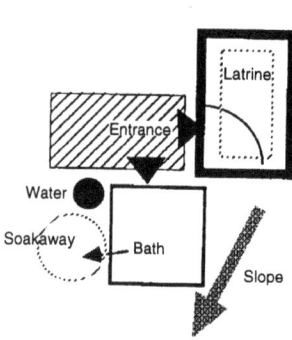

Alternative layouts for the latrine and bath shelters, depending on the lie of the ground: water and wet activities should be kept further down the slope than the latrine

---
2 In Muslim cultures, the way you face while using the latrine may be very important.
3 Sullage is wastewater from washing and other domestic uses.

this may cause the latrine to fill with water or even a total collapse.

Material
Permanent building materials are normally far more expensive than the locally available materials which are traditionally used. Most latrines are more or less temporary constructions, so permanent materials should be used in a way that can be reused when the latrine is full and needs replacement.

Make an estimate of what materials you will need, and ensure that they will be available when you need them. In many cases it is wise to buy the materials in advance, to be sure they are available when you start building. If you are paying for labour it can become very expensive to have people waiting for materials which arrive days or weeks late.

The pit

Round or rectangular pit?
A round pit is more stable than a rectangular one but may be more complicated to cover. Lined pits should as a rule be made round. Latrine builders in stable soil areas seem to prefer rectangular pits. They are easy to cover and easy to climb in.

Climbing in narrow round and rectangular pits is easy if you make holes for the feet in the pit sides. If the pit is wide or the soil too soft you may need a ladder.

Lined or unlined pit?
In loose sandy soils the pit needs to be lined. Lined pits should, as a rule, be made round and can normally be made wider than unlined pits as the cover slab can be supported on the stable lining rather than on relatively weak soil.

Pit depth, durability and security
The durability of a pit depends on its volume, the number of people who use the latrine and the way in which they use it, what type of anal cleaning material is used, how much garbage is put in, etc. It is therefore very difficult to predict the lifetime of any latrine.

For private latrines you may make a rough estimate based on 60 litres per person per year. The true figures, however, may vary from 15 litres up to 300 litres per per-

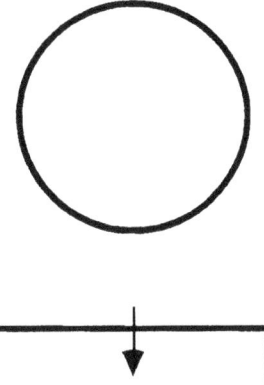

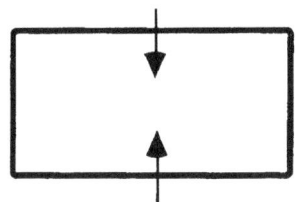

A round pit is stronger than a rectangular pit as the straight sides collapse more easily

son per year. For public and institutional latrines an estimate should be based on experiences from previous latrines under similar conditions.

Do not dig deeper than is safe for the kind of soil you are digging in and remember that large rectangular pits are more liable to cave in than narrow round ones.

Foundation for the cover slab?
To prevent surface streams and pools of stagnant water penetrating under the cover slab, a slight elevation of the slab in relation to the surrounding ground is recommended. With traditional latrines this usually happens automatically, as the logs and soil cover have a reasonable thickness.

For concrete slabs, a foundation to raise the slab above ground level is recommended. The foundation will also prevent excavated soil from falling into the pit. Excavated soil should be used as backfill around the slab and the foundation to prevent any water from streams or pools of stagnant water entering the latrine.

Before you start digging
If you intend to cast the concrete slab yourself, do this before you dig the pit, as the concrete needs to be undisturbed and wet (to cure) for one week before it can be put over the pit. The concrete slab will attain its full strength after four weeks. Until then you should handle it with care and avoid loading it. The casting and curing of the slab will be described in Chapter 5.

When you start digging, any drawings on the ground will disappear. The careful builder will fix the dimensions marked out on the ground with strings and pegs outside the building area. When placing the pegs, remember that the excavated soil will cover a considerable area around the pit, so do not make them too small.

Building a foundation
Assuming that you do not need a full pit lining, you should start by removing the topsoil until you reach firm and undisturbed soil stable enough to carry the weight of the latrine. Laying the foundation will be easier if the bottom is horizontal. If not, you can adjust it with mortar and scrap bricks.

You can now start laying out the first course of bricks in the right place, but without using mortar. Normally you should lay the bricks with the short end facing the

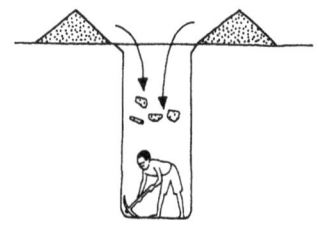

Moving the excavated soil away from the side reduces the risk of accidents

pit. Check your dimensions with the size of the cover slab to make sure that there is just enough support for the slab (3–4 inches or 10cm would be enough). The advanced latrine builder may choose to make the foundation wider and step by step close up the courses. This will give a wider pit which lasts longer.

Continue laying the foundation until it is one foot (30cm) above the ground. Check that the courses are well levelled. If they are not, the whole latrine floor will become tilted. If you want it to be sloping, it should slope gently towards the door, but this is easier to adjust when you put the cover slab in place.

If brickwork is too expensive the foundation can also be made out of mud blocks laid in mud mortar.

Digging the pit
Allow the brickwork to cure for a day or two before you continue digging inside the foundation. You will discover that the foundation is an excellent base to work from, preventing excavated soil from falling down over the person digging in the pit, and is a good platform for the person above in charge of the rope with the bucket.

Remove the top soil

Build the foundation

Climbing in the pit
Leaving the pit may be difficult if the pit is deep. In a narrow pit you may have to make holes in the side walls for climbing. If this is not possible because of the pit size or the strength of the soil you may need a ladder. A good ladder may be made of two long poles (e.g. Eucalyptus or Bluegum) with cross-pieces nailed to the poles. The cross-pieces should be cut out at the ends to fit better to the poles they are nailed to.

Continue digging when the brickwork is strong enough

Pit lining

Bricks and concrete blocks are excellent for lining but natural stones can also be used. Wooden materials may have a limited lifetime but may be all that is available.

Various types of pit linings are described in Chapter 4.

The latrine slab

Traditional slab
A traditional slab is normally made of logs covered with soil. It has a hole (called the drop-hole) to allow use of the latrine. If the soil is loose the logs may be covered

with scrap plastic, old iron sheets, banana leaves or grass. Holes between the logs are sometimes filled with stones or mud before they are covered with soil. Termite soil is frequently used as a top layer, as it becomes hard and is relatively easy to keep clean.

When the latrine is complete a SanPlat can be installed and the surrounding floor given a final covering (if required).

Concrete slab

A concrete slab is in most cases cast by the side of the pit and put in place when the concrete has cured. It can also be cast in casting yards and transported to the site.

The casting of concrete slabs is described in Chapters 5–7. Slabs can be cast with integrated SanPlat features (foot-rests, keyhole-shaped drop-hole, etc).

A traditional pit cover may be used as a mould for casting a slab over a pit.

Brick vaults and domes

A possible way to cover the pit in a permanent way is by making brick vaults for rectangular pits or domes for round pits. This method has been used for over 2000 years in, for example, churches and bridges. If you are interested in trying this method, refer to specialist literature on bricklaying.

A simple method of making excellent concrete domes with SanPlat features is described in Chapter 5.

The superstructure

Privacy and protection

The principal reason for having a superstructure is to provide shelter and privacy for the user. It also protects the pit cover from rain. This is important for latrines covered with mud on poles, especially in areas with heavy rainfalls and unstable soil.

The superstructure is also the visible part of the latrine which means that aesthetic factors are important if the latrine is to be appreciated.

Dimensions and workmanship

Normally the visual aspect of the latrine is inferior to the dwelling house, both in dimensions (height of the door and the ceiling) and workmanship. This may reflect people's attitudes to their latrines. Proper dimensions

The appearance of the superstructure is important if the latrine is to be appreciated

and good workmanship are consequently strongly recommended.

Walls
Walls give privacy to the user. They also carry the load of the roof to the foundations and the ground.

Openings
Daylight is important for hygiene and cleanliness. A minimum requirement is that people can easily see if there is faecal matter on the floor, so as not to step on it. To encourage general cleanliness as much light as possible in the latrine is recommended.

For the latrine to be pleasant to use it should also be well ventilated. At least two windows are recommended to allow air to pass through even when the door is closed. If this is not possible, an opening can be arranged above or below the door.

Doors
Doors can be made in many ways. Square timber is a common solution. Traditional grass or bamboo doors will however fill the same function. In many places L-shaped entrances form a substitute for a door.

Roof
Traditional grass roofs are the most common solution in rural areas while latrines in urban fringe areas tend to have sheet metal roofs.

Many latrines have no roofs at all. There are even situations where people who are not used to using latrines would prefer to have a latrine without a roof rather than defecate 'inside a house'.

4. Common latrine building problems and solutions

The objective of this chapter is to highlight a number of common problems and their possible solutions.

The format chosen is called a PRS-pattern. PRS stands for:

- Problem
- Reality
- Solution

which together constitute a PATTERN of problem solving.

It is advisable to present the solution to each problem on a PRS-sheet with a title which may focus on the problem, the reality or the solution. This helps you to find your PRS-sheet.

You may choose to format your PRS-sheet in any way you like, beginning with either the problem or the solution. It should, however, always include a description of the reality for which it is relevant.

If the patterns are presented in a loose-leaf format, the sheets can easily be sorted for different purposes.

Collapsing latrine pits

Problem
Unstable soils

Reality
Two of the most common reasons for not building latrines are collapsing soil and high water-tables, often in combination with each other.

Solutions
For unstable soils the following measures can be taken:

- Use round pits
- Introduce pit linings
- Reduce pit dimensions.

1. Round pits
In areas where there is a tradition of building latrines the population have already discovered that round pits are more stable than rectangular ones.

In Malawi, for example, where the soil is generally stable, the population as a rule dig rectangular pits. An exception is at the lake shore where pits, as a rule, are round because the sandy soil is less stable. The same is true in Mozambique: in the areas along the coast, the round pit is the only shape you will find, whereas inland people dig rectangular pits. You will find the same phenomenon in most places in Eastern and Southern Africa.

When asking why this is the case, few people refer to soil stability, but instead give tradition as a reason. In areas with mixed soil conditions, where the soil is less stable, there are often no latrines.

It may be tempting to propose round pits for all situations as strength is an advantage and the cost seems to be the same. Covering a narrow rectangular pit is, however, simpler, especially if the availability of long, strong logs is a problem.

The same kind of logic also works for latrine slabs of concrete. Round slabs are generally cast on-site, while rectangular slabs can easily be segmented and transported in pieces.

2. Introduce pit linings

Round pit linings are very stable and can be made relatively cheaply, while rectangular pit linings are structurally unsound and need to be made with strong building materials. Pit linings should therefore normally be made round. The material used will depend on what is available and affordable.

LINING WITH BURNT BRICKS

Burnt bricks, when available and affordable, are very good for pit-lining. Cement or mud mortar in horizontal joints are recommended as the bricks may break under the load of the slab and the superstructure.

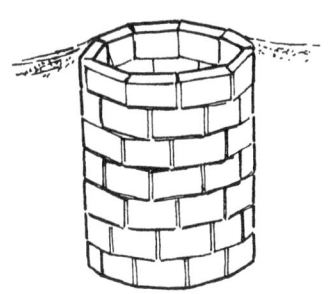

LINING WITH CONCRETE BLOCKS

Due to the production process, concrete blocks usually have exact dimensions. They can therefore be used without mortar, being simply stacked along the sides of the pit. The top courses should be fixed with cement mortar as the surrounding soil is often washed away with heavy rains.

LINING WITH STONES

Natural stone makes an excellent lining material which requires no mortar except in the top layers. The strength of the stone allows it to take a considerable load without breaking, while the weight of the structure holds the stone in place. The top of the lining may need to be set in mortar as the stones might otherwise start falling away. If you are going to use a concrete slab you may need some mortar anyway to fill the gap between the slab and the lining.

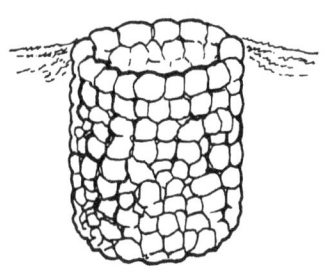

LINING WITH CONCRETE RINGS

Using concrete rings for pit-lining can be very expensive if normal well rings are used. They are heavy to transport and difficult to set in place. Using segmented rings with interlocking joints may be easier.

LINING WITH OIL DRUMS

Old oil drums are frequently used for pit-lining. After the bottom of each drum has been cut out, they can be placed one on top of another to line deep pits.

The small volume and the high cost of the drums make this method fairly expensive. If the sides of the drums are not provided with sufficient perforations, absorption of liquids can be a problem. Another problem is that the sheet metal will eventually corrode and collapse if the drums are too old.

The pit contents and the resultant gases in a latrine are very corrosive.

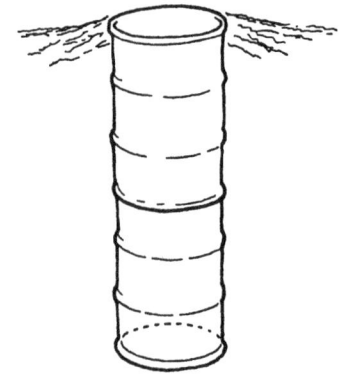

LINING WITH WOODEN BASKETS

Large baskets made of sticks and twigs can be put into the pit. The expected lifetime is around two years but this depends on the material used and whether there are termites.

If the back-fill around the pit is made with well-compacted termite clay, a very stable pit can be achieved in loose sandy soils. When dry, this clay will form a firm pit lining even after the wood in the basket has been eaten by termites.

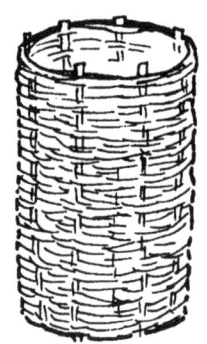

3. Reduce pit dimensions

Apart from the shape, size also influences pit stability, and the risk of collapse. Pits with small diameters are more stable than pits with large diameters.

In areas of unstable soil it may be tempting to compensate for the poor depth by making pits wider to achieve a reasonable volume. This is only possible if the pit is lined. Lining material may be difficult to find or too expensive to use. In such cases the solution is to reduce the pit diameter.

Pits in unstable soils should not be made too deep as a pit collapse during the excavation may have serious consequences for the people digging. It should be possible to dig a 2m pit with relative safety for the builders. Should the soil collapse it would only fill half of the pit, and the person at the bottom would only be covered up to waist level. This should allow the person to breathe while help is organized.

Splashing latrines

Problem
High water-level in the latrine

Reality
In latrines with too much water, unpleasant splashing may be a problem. It is the area of free water which allows the water to splash. The problem is especially common in areas of high water-tables.

Solutions
1. Floating matter
Covering the surface with floating matter will prevent the water from splashing. Polystyrene beads can be used for this purpose. They can be obtained from factories producing plastic items in expanded polystyrene (a very light, white plastic material, commonly used for packing fragile items). Just a slight cover is required to stop the splashing. If these are not available, any floating material will help as long as the surface is reasonably well covered (test with a stone).

In splashing latrines, mosquito breeding may also be a problem that can be solved by a cover of floating polystyrene beads.

2. A sloping surface
A sloping surface under the drop-hole will prevent splashes from reaching the person using the latrine. This surface can be made by using a piece of polythene to form a slide under the drop-hole.

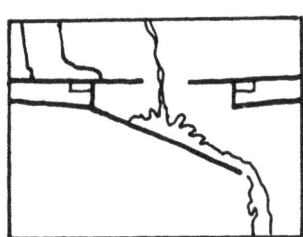
A sloping surface under the drop-hole

3. A twig from a tree
Splashing latrines are often very shallow ones in high groundwater areas.

Catching the faeces before they reach the surface will prevent the splashing. To do this one or two twigs may be stuck into the ground at the bottom of the pit to catch the faecal matter before it hits the water.

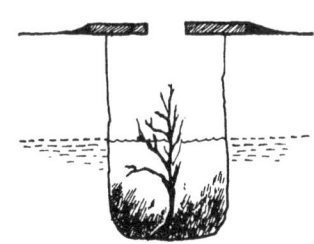
A twig in the pit

Flies breeding in latrines

Problem
VIP latrines and pour-flush latrines are the common solutions to the problem of flies breeding in pit latrines. They are, however, inaccessible to many families.

Reality
Latrines are common places for blow flies[1] to develop. A pregnant female fly will look for a dark place to lay her eggs. After a few days the eggs hatch out as maggots (which are the larvae of the fly). The maggots will live in the latrine pit, feeding on the fresh faeces.

The larvae can easily be seen through the drop-hole if you use a torch light. You can normally see them moving in the bottom of the latrine right under the drop-hole. In severe cases they may form a living mat covering part of the bottom of the latrine. When the maggot is old enough it will build itself a shell (pupa) where it transforms into a fly. The new fly will leave the pit through the drop-hole.

Fly breeding is very irregular and may at times be a nuisance and cause health problems to people by contaminating food with small amounts of faecal matter carried from the latrine pit and from contaminated garbage.

Solutions
Correctly built VIP latrines and pour-flush latrines have shown their value in reducing fly circulation in latrines. For people who cannot afford such latrines, several simple solutions are available.

1. Hot ashes
The fly maggots are sensitive to heat and ashes. Hot ashes thrown into the latrine have proven to be an effective way to reduce fly breeding. The heat will kill most of the maggots, which usually appear directly under the drop-hole to feed on fresh faeces. It is suspected that cockroaches, which also live in the latrine pit, may kill the remaining larvae, especially in very dry latrines where the wet spot under the drop-hole is limited.

---

[1] The fat blow fly or 'bluebottle' is easily recognized by the colour of the back which is metallic blue or green.

2. Hot water
A bucket of almost boiling water should kill a large population of maggots if the infestation is more serious. This method should not, however, be used too often as the fly maggots also like humidity in the pit.

3. Biological control
Bacillus thuringiensis bacteria[2] attack young fly larvae.

---

[2] This is available commercially through: Oy G.A.C. Products Ab, Sportvagen 8, FA2, SK-02700 Grankulla, Finland.

Mosquitoes breeding in latrines

Problem
Mosquitoes may transmit disease

Reality
Wet latrines may become a breeding place for mosquitoes. The type of mosquito that breeds in latrines does not cause malaria. It can, however, transmit urban filariasis ('elephant leg') if this disease is common in the area.

Solution
Mosquito breeding stops when a cover of some sort develops on top of the water. This is because the larvae need a free water surface in order to breath. In latrines with mosquitoes, breeding can effectively be stopped by adding one inch of polystyrene beads to the latrine. The use of polystyrene beads is also described earlier in the chapter in the section 'Splashing latrines'.

A separate urinal

Problems
Dirty squatting plates
Fly breeding
Fast filling up

Reality
Men have difficulties urinating in a latrine hole. In a traditional unimproved latrine, urine may even destroy the floor around the drop-hole. Fly maggots like water and humidity, and seem to have nothing against the acid urine. Urine also retards biological decomposition of faecal matter in latrines (too much nitrogen). Urine is also a good fertilizer which can be used directly if diluted with water.

There are consequently several good reasons for making a separate urinal close to the latrine, especially if it has a traditional mud floor.

Solution
In southern Mozambique, latrines have no roofs, but are built as fenced-off corners of the plot.

A separate urinal can easily be made by filling a pit with stones to make a soak away. An old clay pot with a hole in the bottom can be used as a top to prevent the urine from splashing. In Bangladesh the traditional shape of the pots is excellent for collection of the urine.

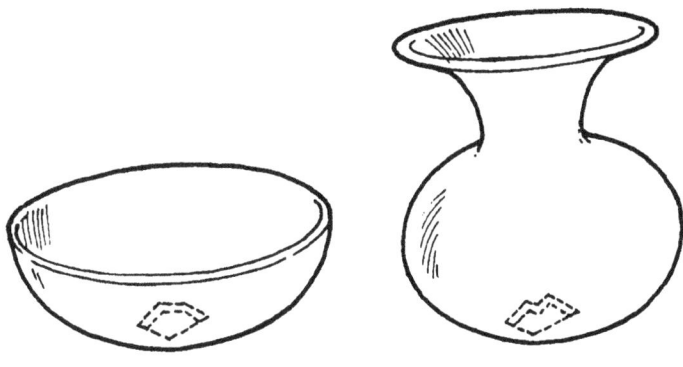

Typical clay pots, each with a hole in the bottom, used as urinals over a hole filled with stones

Dirty latrines and hookworm breeding

Problem
Traditional latrines are difficult to keep clean. Faecal droppings may stick to the mud providing a breeding ground for parasites (hookworm).

Reality
The hookworm lives inside the human body. Its eggs develop in humid soil and penetrate bare skin, usually the feet. For obvious reasons the hookworm is very common in developing countries and seriously affects the health of people as it feeds on blood.

Solutions
1. The use of shoes is normally recommended against hookworm as the worm cannot penetrate shoe soles.
2. Use of latrines with SanPlats. As it is isolated well inside the latrine, the hookworm cannot reach the user. If the latrine has a concrete slab (SanPlat) the worm's life cycle is terminated.

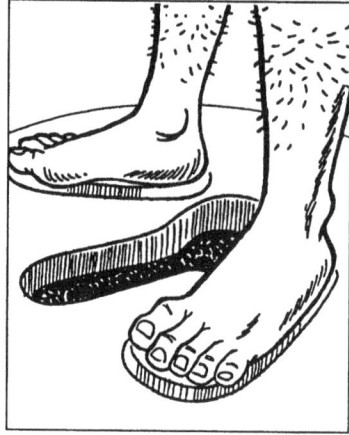

If you do not like where you have to put your feet, try using a SanPlat

Collapsing pit covers

Problem
Collapsing pit covers in traditional latrines

Reality
Wood in contact with soil will eventually be attacked by rot and/or termites. Some species of wood are more resistant than others. Another reason for collapsing pit covers may be the use of logs which are too short to span the pit properly.

Solutions
1. Thicker or more resistant logs
If the reason is termite attacks, you could try thicker logs or a more resistant species of wood. Thicker logs, having more heart wood, normally resist both rot and termite attacks better. Certain species, such as some palm trees, are very resistant and will last for 30 years or more.

2. Longer logs
If the reason is soil collapsing at the top of the pit longer logs may be a solution.

3. Brick pit collar
If longer logs are difficult to get, a pit collar of bricks may be used.

4. Ring beam
Casting a ring beam in the soil before digging will reinforce the soil at the top of the pit. As always, the round pit is stronger than the rectangular one.

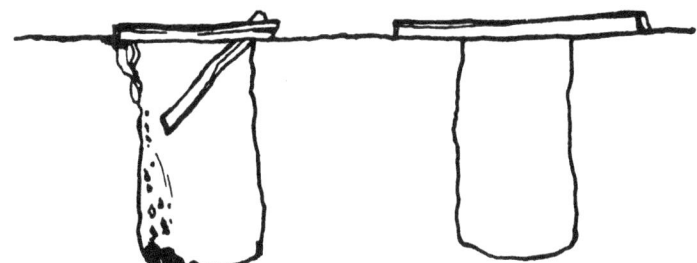

Logs which are too short may result in soil collapse at the edge of the pit, especially if the soil becomes wet

When the latrine fills up

Problem
The latrine is full.

Reality
All latrines eventually fill up. Replacement may be a problem in terms of space or cost.

Solutions
1. Build a new latrine
The most common solution is to build a new latrine.

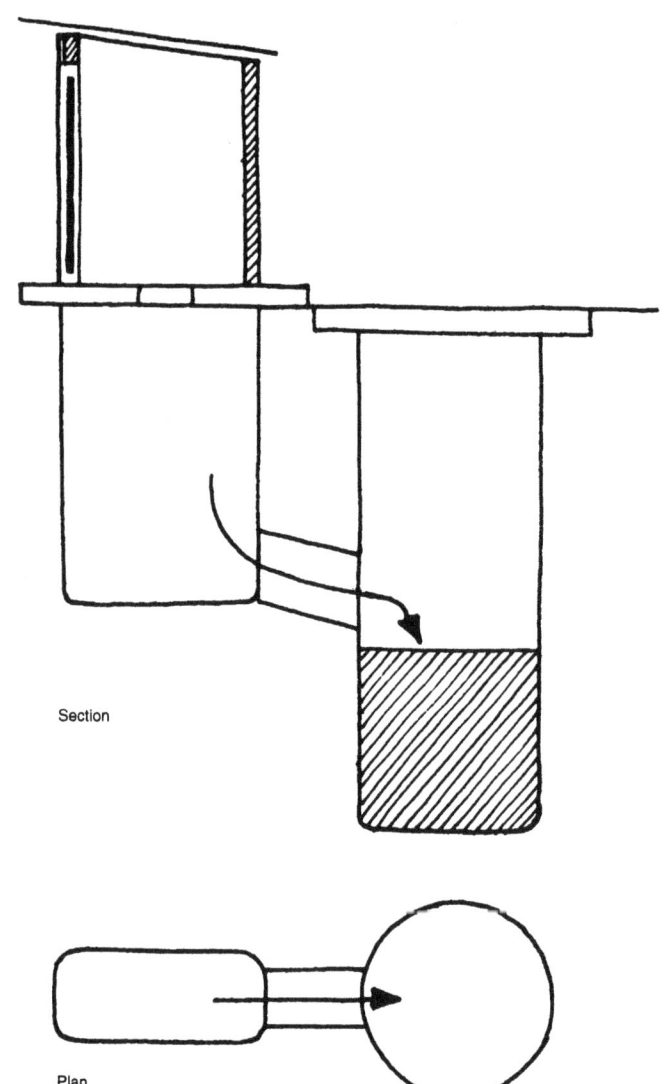

Two pits can be used for ever if the old pit is dug up when the other is full; after a two-year interval, it is completely safe to do so

Reuse as much as possible of the building material; for example, the SanPlat.

2. Emptying into a temporary pit
If space is a problem, latrines can actually be emptied. If the soil is very stable a deeper pit can be dug by the side of the old one and an opening made, allowing the pit contents to flow through into it.

This solution is hazardous, but has been practised in Lamu on the Kenyan coast for hundreds of years. In Tanzania it is called the 'frogman' solution.

3. Emptying with a cesspool cleaner
In urban areas it is actually possible to empty many latrines with vacuum-tankers (cesspool cleaners). To make the pit contents fluid enough to enter the hose and be sucked up, you may need to add some water and stir it using a pole with an extra piece of wood nailed to the top. A few pails full of water will work wonders if you stir it up (5–10 per cent water).

Solid objects and plastic bags dropped into the latrine can be a serious problem as these may block the hose of the vacuum-tanker.

Note: There is a risk of the whole latrine collapsing when emptied. An extra hole next to the pit (diagonally, at one end of a rectangular pit) may be a safer solution than trying to empty it through the drop-hole. The hole may be repaired with some stones and wet mud when the pit is emptied.

When space is a problem

Problem
Finding the space to build a new latrine

Reality
Space can be a problem especially in slum and squatter areas. Space problems can include the following:

1. Many latrines and garbage pits have been built in the same area and now there are no undisturbed sites left.

2. There is a shortage of land. Houses are built so close together that there is no room to build a new latrine.

All latrines eventually get filled up. Replacement may be a problem in terms of space or in terms of economy.

Solutions
1. Reuse of space
Build a new latrine on the site of an old pit. After two years it is not a health hazard to dig up an old latrine.

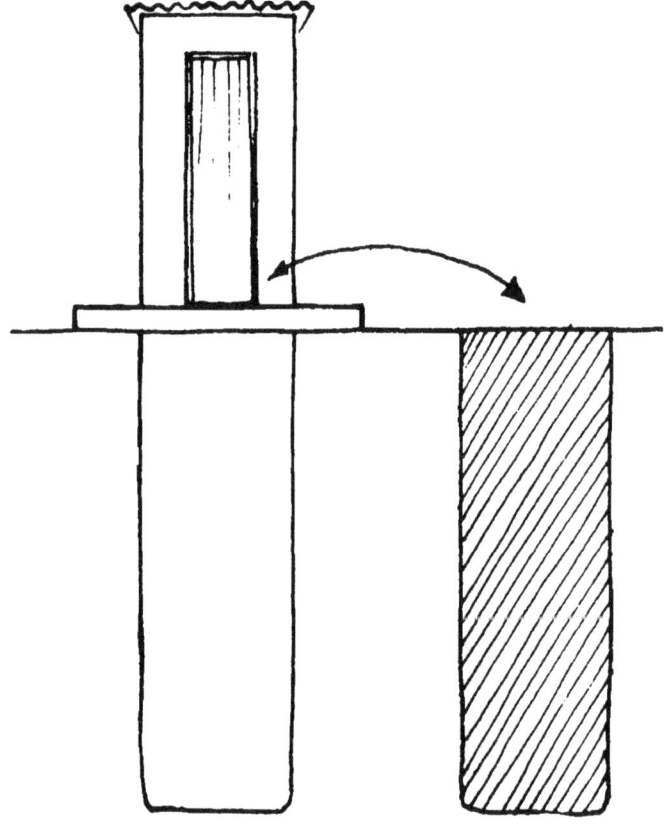

Reuse as much as possible of the building material from the latrine which is full and cover the pit for other uses.

2. Pot-and-bucket latrines with lid

The bucket latrine has a bad reputation for being insanitary. A social stigma is sometimes also attached to the people emptying them. In terms of space, however, they are very economic, as the bucket with a tight-fitting lid can be kept anywhere and the chamber pot has its given place under the bed.

A piece of paper in the chamber pot or some water makes it easier to clean.

An old magazine provides anal cleaning material while at the same time it can serve as a lid to the chamber pot.

Roofs

Problem
Should the project promote roofed latrines?

Reality
The roof is generally appreciated by the users. It protects not only the user but also the slab from being wet, which may be important for the durability of the latrine.

From a hygienic point of view sun, rain and wind on a concrete slab may be very positive, especially if the surfaces are smooth and slightly sloping as this will keep the latrine slab cleaner.

Some (older) people who were educated to defecate outdoors feel very reluctant to do so 'inside a house', which they would prefer to use for other less smelly purposes.

The roof is a considerable part of the cost, especially in areas where natural thatching material is scarce or inappropriate to use.

Solution
Discuss the possibility of roofing or not roofing the latrines with staff members and local leaders. In many cases it may be wise to leave it up to the individual families to decide whether they want their latrines roofed or not.

A roof may add significantly to the cost of latrines and be unacceptable to some people

Windows and openings

Problem
Should latrines have windows and other openings?

Reality
Light and fresh air in the latrines are always appreciated by the users. Few people, if any, appreciate a dark latrine, especially if they have bare feet and risk stepping in somebody's faeces.

In rural West Africa, many people fear there may be demons in dark latrines.

Light in latrines may attract flies. A lid over the drop-hole deters the flies.

VIP latrines are normally recommended to be dark and have no lids in order to control the circulation of flies. Practical experience indicates that VIP latrines work excellently with a lid over the drop-hole.

Solution
Latrines should have good light and good ventilation. For this at least two openings are recommended to allow for cross ventilation.

A lid over the drop-hole will give the required darkness in the pit.

Note: Lids are not recommended in public and institutional latrines as the handles will become filthy. VIP latrines or pour-flush latrines are recommended in such situations.

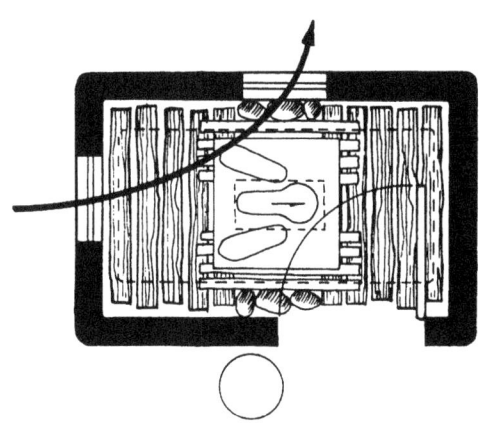

Light and fresh air in latrines are always appreciated by the users

Subsidies for different types of latrine

Problem
Should latrines with different designs have different subsidies?

Reality
Commonly VIP latrines or any other more sophisticated latrine type are already being subsidized by the programme at the time when, for example, SanPlat latrines are introduced.

A common policy is for projects to provide cement, reinforcement and screening material while the individual builders have to provide sand, gravel, bricks and other local materials. Considerable quantities of cement etc. are provided to each family building a new latrine.

Applying the same policy for SanPlats will put the project in a situation where people who can afford the more sophisticated latrines will receive large subsidies, while the SanPlat families receive very little, a solution which is obviously unfair and does not promote the more cost-effective solution.

Solution
The obvious solution is that each family should receive equal subsidies if they improve their latrine to a level which corresponds to the minimum requirement of the project.

A consequence of this may be that the subsidy can be reduced and benefit more families.

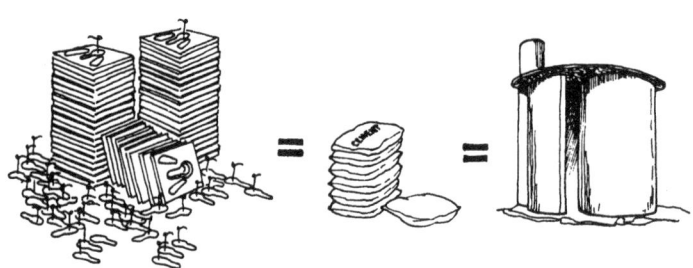

Many families can be given SanPlats free of charge for the same cost as all the cement required for one VIP latrine.

5. SanPlat making

This chapter explains step by step how to make different SanPlats.

How to make small SanPlats 60×60cm

SanPlats can be made in different shapes and sizes. This chapter explains, step by step, how square, flat SanPlats and dome-shaped SanPlats are made. The method for making them is practically the same.[1]

Moulds

The following moulds are required:

- SanPlat frame (one per slab and day)
- Drop-hole moulds (one per slab and day)
- Foot-rest mould (one or two should be enough)

Manufacture of the moulds is described in Chapter 6.

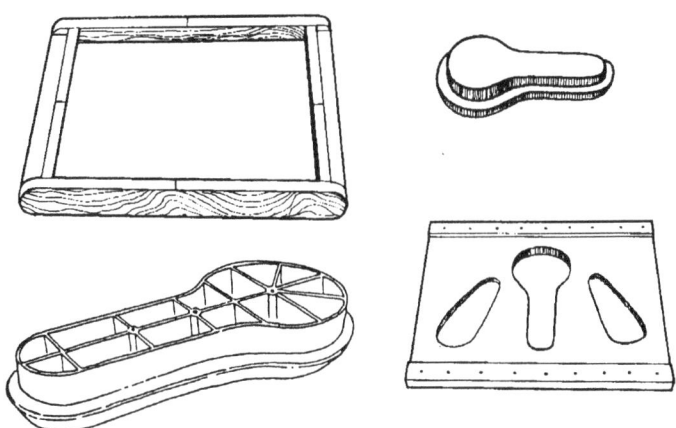

Bottom left: Drop-hole mould in plastic, made by LCS Promotion

---

[1] Small SanPlats (60×60cm) can also be made using full-size all-in-one plastic moulds available from UNICEF, Freeport, Copenhagen, Denmark or directly from the producer, LCS Promotion, Flo 18, S-46796, Grästorp, Sweden. (See Appendix 2 for method.)

Tools

A set of normal mason's tools is a minimum requirement. The number of tools needed depends on the daily production.

- One or two shovels or a hoe for mixing the concrete
- Mason's trowels, one large and one small
- A steel-float (floor trowel) for finishing off the surface
- A wheelbarrow for transport of material
- One or two buckets for measuring the material
- A hammer for various purposes
- A hacksaw or a chisel for cutting the reinforcement bars (they can also be cut with a hammer against the edge of a pick-axe or any other sharp edge)
- A piece of a water pipe is useful when bending reinforcement bars for the handles.

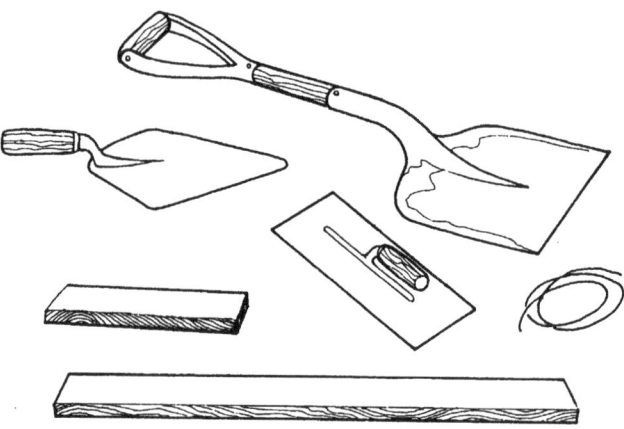

Tools for traditional SanPlat making

Material

The SanPlat slab is made of unreinforced concrete. To avoid unnecessary breakages the material should be clean and of good quality.

The quantities will depend on the dimensions. You will need:

- Normal cement (standard Portland)
- River sand
- Gravel (12mm or similar)
- Plaster sand (if cheaper than river sand) can be used for the moulding. Plaster sand can also be used to 'modify' the river sand if it is too coarse.
- 4–6mm mild steel reinforcement (for the handle of the lid).

The casting yard

Before you start, make an assessment of the area and the use of the space in your casting yard. You will need a flat, hard and smooth surface for mixing the concrete and later you will need space for spreading out the slabs and for making foot-rests and lids. Finally you may need some space for storing finished slabs. A layout for a typical casting yard is outlined below.

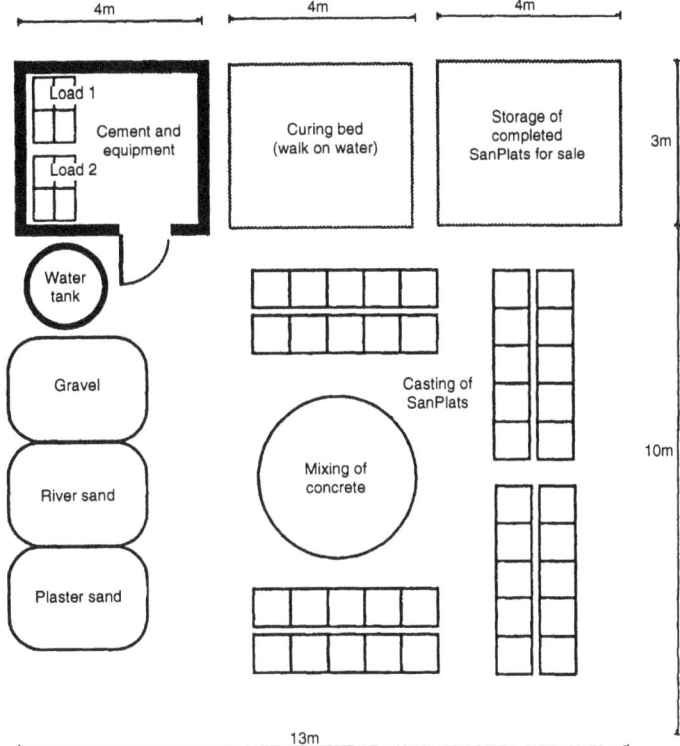

Layout of a typical casting yard for the production of small SanPlats

Making the slab

Good workmanship is always appreciated. It is also a quality which is important for hygiene and cleaning, as people prefer to care for a SanPlat which is smooth and well shaped. No part of any surface must be rougher than the surface of a well-stretched-out palm of a hand and the shape should always be the same as that of the moulds.

Keep the moulds clean. Remove concrete leftovers before they get too hard.

1. Start by arranging a flat area using a straight piece of wood and put the frame on it with the narrow side up (sides should be slightly tilted).

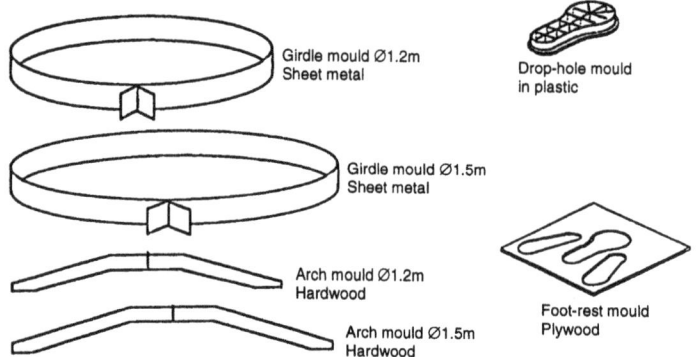

Keep moulds clean and remove cement as soon as possible, preferably while it is still fresh

2. Cover the bottom with paper from cement bags or newspaper. If necessary, fix the papers temporarily with stones or cement lumps so they do not blow away.

3. Place the drop-hole mould in the middle and check its position with a string (there should be centre-marks in the frame as well as in the drop-hole mould).

The moulds are put on a flat surface and the drop-hole mould centred using a string

4. You are now ready to cast the slab with concrete in the proportions 1+2+2 of cement, sand and half-inch (12mm) or smaller clean gravel. Start by filling very carefully around the drop-hole and pound with the edge of the trowel to make sure that the concrete fills well and you get a smooth edge around the hole.

5. Spread out the concrete over the paper with the help of a straight batten to ensure correct thickness of the concrete.

6. The slope towards the drop-hole is made by supporting the batten on the edge of the side of the drop-hole mould.

The reinforcement should be placed exactly in the middle

The correct inclination inwards is produced using a batten

7. Now the surface of the slab can be worked out to final finish, preferably using a steel-float (floor trowel). You may need to wait for some of the water to be soaked up by the underlying paper and the sand in order to get a good surface. Some cement powder sprinkled on the surface may also be helpful in achieving this.

8. After an hour or two, when the concrete is stiff, the drop-hole mould can be removed. This is done more easily if you hit the two ends of the mould gently with a hammer.

9. This is also the right time to make the inscriptions in the slab. Each slab should have the concrete mix, the casting date, the number of the slab and the initials of the slab maker written on the concrete as indicated in the illustration on page 66. (The numbers of the slabs should be consecutive numbers starting from 1, independent of the size and shape of the SanPlat.)

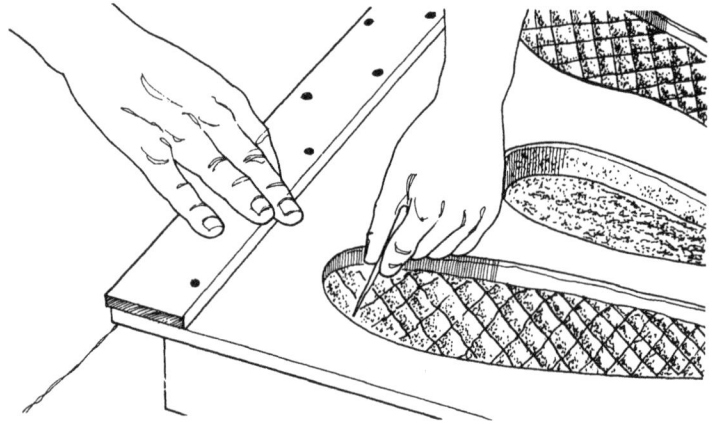

Use the foot-rest mould to score the area for the foot-rests

10. Before the concrete becomes too hard you should scratch the surfaces of the foot-rests to get a rough surface for the foot-rests themselves to stick better. Use the foot-rest mould as a template.

Making the lid

To guarantee a good fit between the slab and the lid (so that no smell and no flies can pass) the lid should be cast in the drop-hole of the slab. The lid is therefore cast in the hole of the slab when the slab is resting on the ground, using the hole as a mould. Lids can also be cast in special moulds available from LCS Promotion.

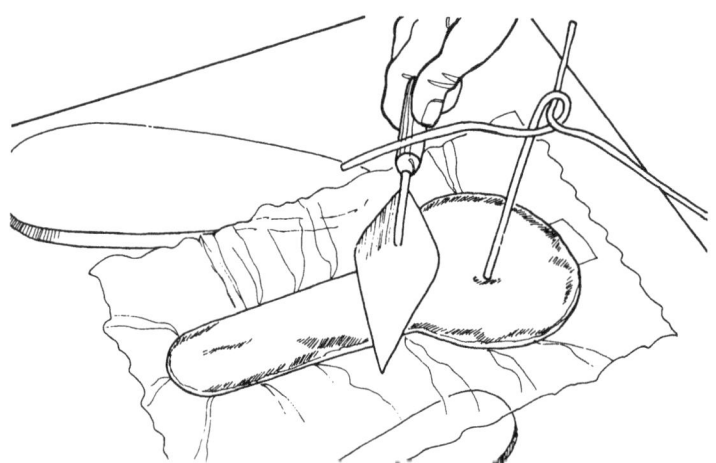

A perfectly fitting lid can be cast using the drop-hole as a mould

1. Start by cutting 750mm and 600mm pieces of a round 6mm reinforcement bar for the handles and reinforcement and bend them as shown in the diagram.

2. Remove any rough concrete edges and put some sand in the drop-hole up to the bottom of the slab. Put in

a wet sheet of paper and fit it carefully to the sides of the hole before half-filling the hole with concrete. If some extra concrete is added a slightly heaped form can be achieved, giving extra strength to the lid.

3. Put the handle and the reinforcement in place with the vertical part of the handle well centred in the middle of the round part of the lid and finish off the surface.

Foot-rests

The foot-rests are cast in the foot-rest mould when the slab is resting on the ground.

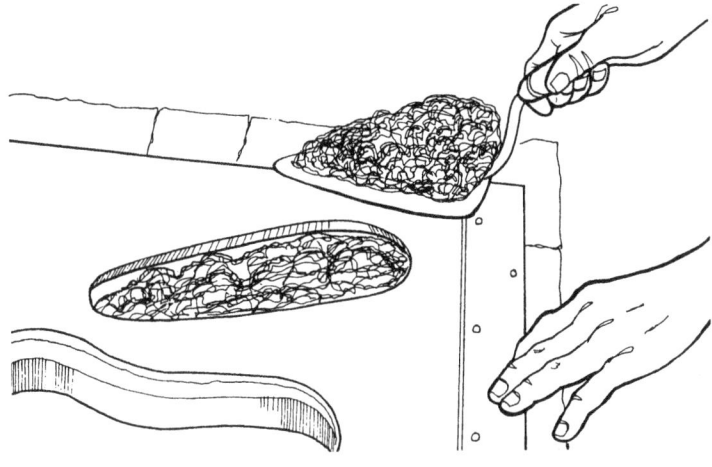

The foot-rests are made by filling the holes in the foot-rest mould with cement mortar

1. Start by wetting the scratched area and rub some cement into the scratched surface for better adhesion.

2. Place the mould in the exact position using the shape of the drop-holes of the slab and the mould for guidance.

3. Fill the foot-rests with concrete and finish off the top surface with a small mason's trowel. A little extra cement on the surface may make it easier to get a smooth finish.

4. Leave the foot-rest mould for a few minutes and allow the cement to harden before removing it very carefully, not disturbing the top surface.

5. Clean the edges and finish them off with the small trowel.

6. The lid and the foot-rest need to be kept wet for curing. The curing time may be reduced to one or two days if extra cement is added to the concrete. (Foot-rests are integrated in the full-size all-in-one SanPlat mould.)

Curing

The curing is best done in a curing tank or in a curing bed as described below. The recommended curing time is always one week.

In a curing tank
For curing in a tank, make sure that the SanPlat is well covered with water. If this is not possible, cover the SanPlat with a plastic (polythene) sheet.

On a curing bed
1. Make a shallow, flat hole in the ground one foot (30cm) deep and as wide as you need. Note that the edges should be horizontal because we are building a shallow water tank.
2. Stabilize the bottom soil with cement 1:20 and make strong edges with bricks. Water and compact well.
3. Build up the sides with bricks (30cm) and plaster the surface with a fine layer of cement mortar 1+1 to seal the surface and allow it to cure for one day. (We now have a flat tank.)
4. Fill with broken stones and water, and allow to cure for another couple of days before you start using it.

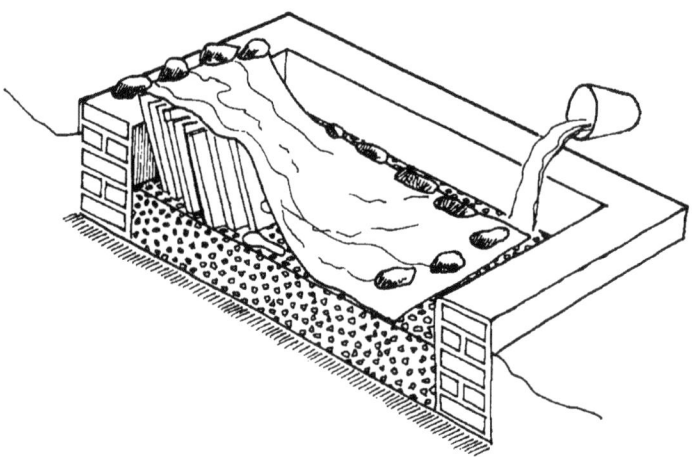

The curing bed should be topped up with water to the level of the broken stones, but no higher

To cure the SanPlats, you simply stack them on the stones, check the water level and cover with a plastic (polythene) sheet. The humid atmosphere under the plastic will provide as much water as the SanPlats will need for the curing.

Note: some plastic is sensitive to direct sun (UV radiation). Choose a UV-protected type or have the area covered with a simple roof. Keep animals away from the plastic.

Testing

When at least one week old, the SanPlats should be test-loaded with the weight of one person, when supported by four diagonally placed wooden wedges in the corners.

How to make dome-shaped SanPlats

SanPlats can be made in different shapes and sizes. The size will depend on the moulds used. The method for making them is, however, the same.

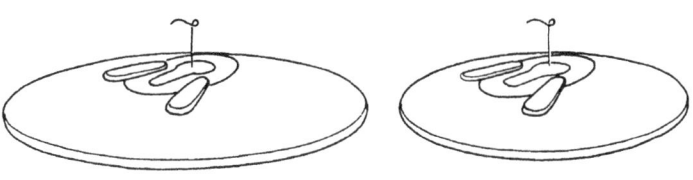

Far left: Circular SanPlat 1.5m diameter; dome-shaped SanPlats are used where wood is in short supply
Near left: Circular SanPlat 1.2m diameter

Moulds

The following moulds are required; the number depends on the daily production:

- Girdle moulds (one per slab and day)
- Drop-hole moulds (one per slab and day)
- Arch moulds (one or two should be enough)
- Foot-rest moulds (one or two should be enough)

Manufacturing methods for the moulds are described in Chapter 6.

Tools

As for the ordinary SanPlat, a set of normal mason's tools is a minimum requirement. The number of tools needed depends on the daily production.

- One or two shovels or a hoe for mixing the concrete
- Mason's trowels, one large and one small
- A steel-float (floor trowel) for finishing off the surface
- A wheelbarrow for transport of material
- One or two buckets for measuring the material
- A hammer for various purposes
- A hacksaw or a chisel for cutting the reinforcement bars (they can also be cut with a hammer against the edge of a pick-axe or any other sharp edge)
- A piece of a water pipe is useful when bending reinforcement bars for the handles.

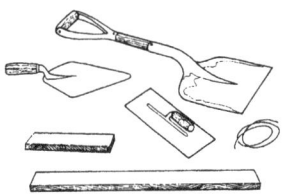

Ordinary mason's tools are used for making SanPlats

Material

The SanPlat slab is made of unreinforced concrete. To avoid unnecessary breakages the material should be clean and of good quality.

The quantities will depend on the dimensions. You will need:

- Normal cement (standard Portland)
- River sand
- Gravel (12mm or similar)
- Plaster sand (if cheaper than river sand) can be used for the moulding. Plaster sand can also be used to 'modify' the river sand if it is too coarse.
- A roll of sisal cord or any other string for tying the ends of the girdle mould together. (Iron wire is not recommended as it will damage the girdle mould when it is pulled with the pliers.)
- 6mm mild steel reinforcement (for the handle of the lid).

Use only fresh and clean materials

The casting yard

Before you start, make an assessment of the area and the use of the space in your casting yard. You will need a flat, hard and smooth surface for mixing the concrete and later you will need space for spreading out the slabs for making foot-rests and lids. Finally you may need some space for storing finished slabs.

The example illustrated is designed for a daily production of five to ten SanPlats of 1.5m diameter. In selecting a site you should try to find a place which people normally pass by. Seeing the slabs being cast makes people curious and is very effective publicity. In planning the

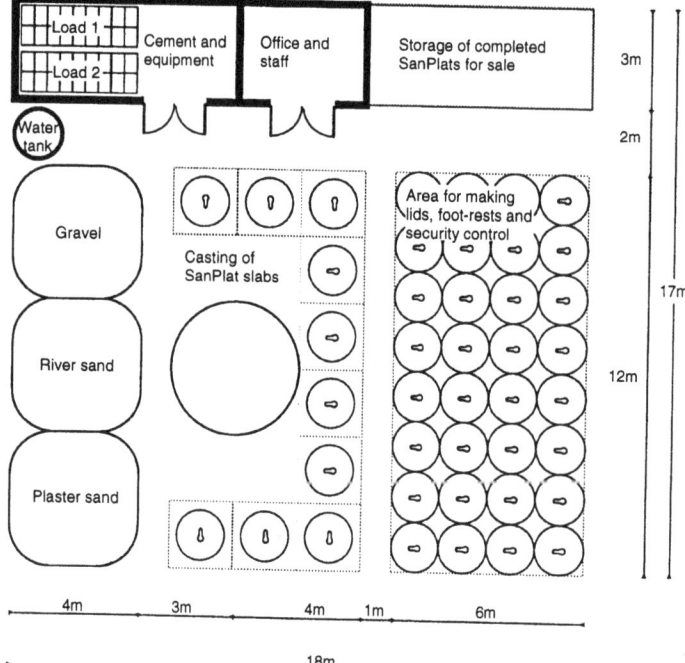

Layout of a typical casting yard for the production of dome-shaped SanPlats

site, you must consider carefully where and how to off-load the site vehicles, as heaps of sand are difficult to move. Pay attention to the vehicle turning radius and to the strength of the road as manoeuvering large heavy vehicles may cause damage. The mixing area should be horizontal and flat. To make cleaning easy, it should be very smooth and have no elevated edges.

Making the slab

Good workmanship is always appreciated. It is also a quality which is important for hygiene and cleaning, as people prefer to care for a SanPlat which is smooth and well shaped. No part of any surface must be rougher than the surface of a well-stretched-out palm of a hand and the shape should always be the same as that of the moulds.

Keep the moulds clean. Remove concrete leftovers before they get too hard.

1. Start by placing the peripheral girdle mould in a circle on the ground and tie the ends together with a piece of string and fill with sand.

2. Place the drop-hole mould in the middle on a heap of damp sand and check its position with the arch mould.

3. Well-positioned, you should be able to turn the arch mould around within the girdle mould.

4. The tips of the arch should move just inside the edge of the girdle mould. Compact the sand and add or remove sand as necessary until you have just the right shape. Be careful that the two wings of the arch mould are resting on sand all the time.

5. Now take the drop-hole mould out of the sand...

6. ...and cover the hole gently with sand without disturbing the shape of the mounded sand.

7. Finally check the height of edge (the thickness of the SanPlat) and remove excess sand as required. The thickness at the edge should be 4cm (1.5in) which can be best checked with a piece of wood of the same thickness.

8. Cover the sand with paper from cement bags or with newspaper starting from the sides and finishing at the top. The papers should cover the sand as tiles cover a roof to prevent concrete later flowing between the papers. Fix the papers temporarily with stone or cement lumps so that they do not blow away.

9. Now place the drop-hole mould in the centre on

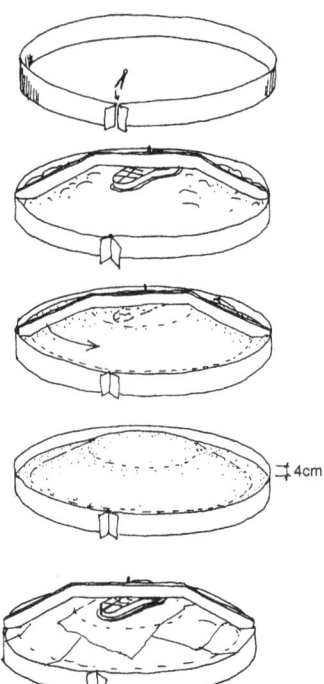

top of the paper and check the position with the arch mould. If another slab should be cast on top it is important that the mould is placed exactly in the middle. If not you will experience problems when rotating the arch mould.

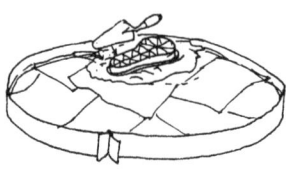

10. You are now ready to cast the slab with concrete in the proportions 1+2+2 of cement, sand, and half-inch (12mm) clean gravel (or finer). Start by filling very carefully around the drop-hole and pound gently with the edge of the trowel to make sure that the concrete fills well and you get a smooth edge around the hole.

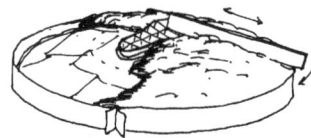

11. Spread out the concrete with the help of a straight batten to ensure correct thickness of the concrete.

12. The inclination inwards around the drop-hole is made by hand 'digging' with a mason's trowel down to the threshold of the drop-hole mould and possibly 1 or 2mm below that level when you finish the surface, to make sure that the edge does not break when you later remove the mould. Pay special attention to the inclination in front of and behind the mould where some extra concrete may be needed.

13. Now the surface of the slab can be worked out to final finish, preferably using a steel-float (floor trowel). You may need to wait for some of the water to be soaked up by the underlying paper and the sand in order to get a good surface. Some cement powder sprinkled on the surface may also be helpful.

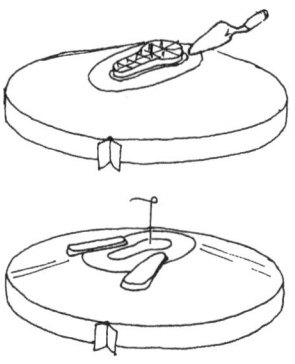

14. After an hour or two, when the concrete is stiff, the drop-hole mould can be removed. This is done more easily if you hit the two ends of the mould gently with a light hammer.

15. This is the right time to make the inscriptions in the slab. Each slab should have the concrete mix, the casting date, the number of the slab and the initials of the slab maker written on the concrete as indicated in the illustration on page 00. (The numbers of the slabs should be consecutive numbers starting from 1, independent of the size and shape of the SanPlat.)

16. Before the concrete becomes too hard you should scratch the surfaces of the foot-rests to get a rough surface for the foot-rests themselves to stick better. Use the foot-rest mould as a template.

Multiple slab making

A number of slabs can be cast on top of each other. This

saves space in the casting yard and facilitates curing as only the top slab will be exposed to the sun, and the moisture from the fresh concrete will prevent the underlying slabs from drying during the curing time.

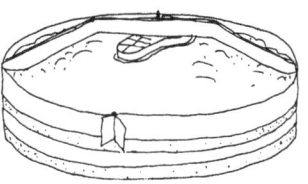

17. To make another slab the girdle mould can be pulled up with a pair of pliers. Before you start to put on more sand as a base for the next slab, place the drop-hole mould in the centre, in the hole of the underlying slab.

18. Now you can continue to make one or more slabs on top of the first one, following the same principles as for the first.

Curing

The curing is best done while the slabs are still stacked. The recommended curing time is one week.

1. Make a hole in the paper through the drop-holes and add water. The sand will soak up the water until the whole stack is well soaked.

2. Even if the top slab will receive some water from beneath, the surface must be protected from drying out. Pull up the girdle mould a bit and cover with more sand and water. Keep the stack wet for one week by watering at least once a day.

3. After one week the slabs can be removed from the stack and put flat on the ground for completion with foot-rests and lids for testing.

Making the lid

The lids and the foot-rests are made in the same way as for the small SanPlats. (See beginning of this chapter for further details.)

Test for safety

When at least one week old, the SanPlats should be test-loaded with the weight of six people, when supported by four diagonally placed wooden wedges as in the diagram. All slabs should be test-loaded. When the slab has cured for at least seven days (check with date inscription) it can be test-loaded with six people on top of it, standing in a row along the length of the slab. Start by checking the date inscription. With a piece of chalk or charcoal, mark out places for a set of wedges to be inserted, first along the length axis and then at right angles. Insert the wedges, checking that all four wedges

provide good support for the slab. Ask people to stand in a row on the slab. If the slab does not break, it has passed the test and should be marked with a small 'S' (for security tested).

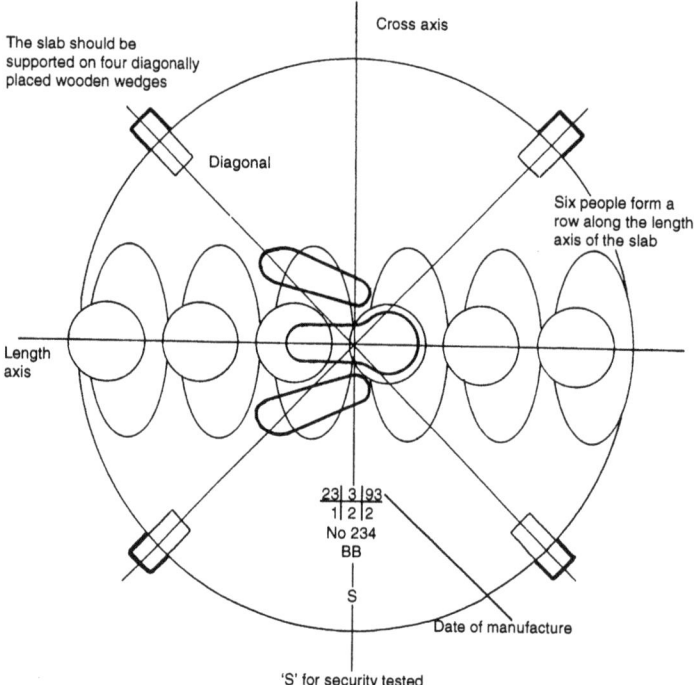

Safety test: the slab should support the weight of six people

6. Moulds for SanPlat making

This chapter describes the different moulds required for making SanPlats and how they can be manufactured.

Types of moulds

SanPlats are normally cast with the top side 'face' up. The process is labour intensive and uses as little material as possible.

All the moulds can be purchased (see page 53), although most of them can easily be produced locally in wood, as described in this chapter.

The 'all-in-one' SanPlat mould, made of plastic

Face-up moulds

Three moulds are normally required for the production of rectangular SanPlats:

- the SanPlat frame
- the drop-hole mould
- the foot-rest mould.

Four moulds are required for the production of dome-shaped SanPlats:

- the girdle mould
- the drop-hole mould
- the arch mould
- the foot-rest mould.

The drop-hole mould and the foot-rest mould are identical for all SanPlats. For design and dimensions please refer to drawings.

Face-down moulds

Only one mould is required:

- the 'all-in-one' mould (described in Appendix 2).

Material

Wooden moulds
The moulds can be made of different materials. A local well-seasoned hardwood has been the most usual choice. This is because the same moulds will be used for making a large number of SanPlats and will be subject to wear and tear from cement, sand, gravel and water, so it is advisable to avoid soft wood.

Sheet metal moulds
The girdle moulds should be made in sheet metal, preferably 1mm thick.

Plastic moulds
Plastic moulds have the advantage of not splintering. The dimensions are exact and they will provide a well-fitting lid. The following moulds are available in plastic:

- The drop-hole mould
- The lid mould
- The 'all-in-one' 60×60cm SanPlat mould

Workmanship

All surfaces of the wooden mould should be planed smooth and well sandpapered.

Curved forms are best made on a spindle moulder which can be adapted to the type of surface required. To avoid splintering of the wood the cutting edges of the mould should be well sharpened and the machine should work at high speed.

Note: The spindle moulder must be operated by authorized trained personnel as accidents can be very serious and even fatal.

Use and maintenance

Always keep your moulds clean.

Cement is easy to remove with water and a brush when fresh, but very difficult to remove if left for a longer time.

Grease or motor oil is not recommended as it makes the moulds and your job dirty and unpleasant. Use water and a soft brush and remove the cement immediately.

For greasing plastic moulds, follow the manufacturer's instructions

The SanPlat frame

The SanPlat frame is composed of six pieces of 20mm planed wood.

Note: Four of the pieces are angled 5mm, so special attention has to be paid when cutting end surfaces.

The pieces are normally joined with 4–inch (100mm) nails (four in each corner). To avoid splitting of the wood, pre-drilling with a slightly smaller diameter drill is recommended.

The drop-hole mould

Wood

The drop-hole mould is made in two parts which are joined together with five 50mm screws. The joint must be strong, as beating on the top part of the mould with a

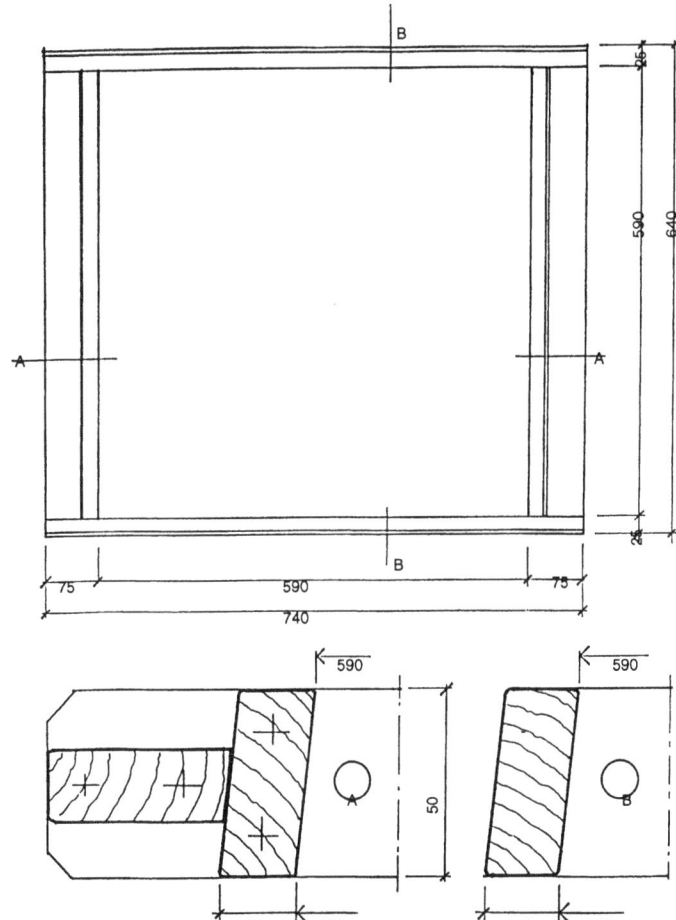

SanPlat frame (material: hardwood; dimensions in millimetres)

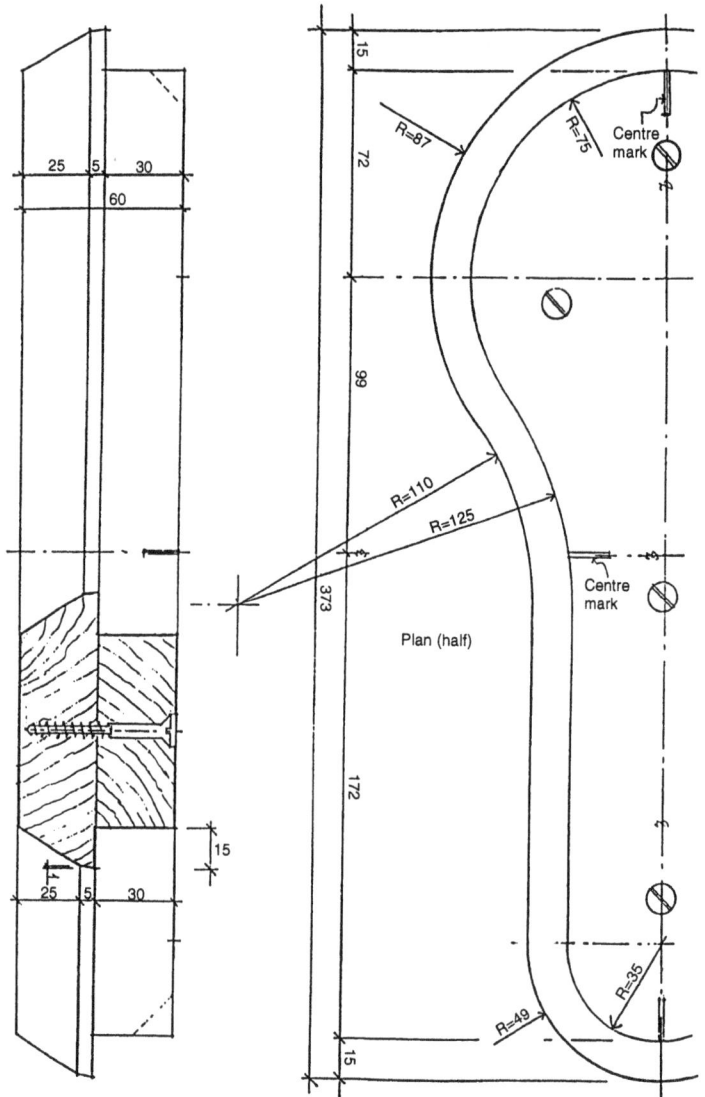

Drop-hole mould (dimensions in millimetres, reduced, not to scale)

hammer before taking it out of the stiff concrete is recommended. It is advisable to make the moulds with very accurate dimensions (maximum error ±1mm) to enable lost lids to be replaced with the same good fit for smell and fly control.

The two parts are best made on a spindle moulder using a guide with the same external dimensions, made of 20mm plywood, which is temporarily fixed to the piece being machined.

Plastic

Drop-hole moulds in plastic have the advantage of being smooth and exactly alike, allowing all lids to be inter-

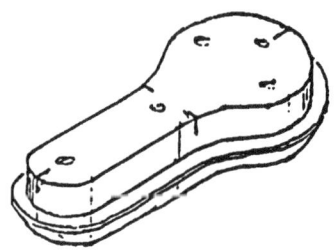

Wooden drop-hole mould

changeable. Confusion with numbers is therefore avoided.

The foot-rest mould

The foot-rest mould can most easily be made in 6–10mm water-resistant plywood. Alternatively, and more commonly, planed hardwood (10mm) is used. The softer superficial parts of the tree are also acceptable as a substantial width (300mm) is required. The soft part of the wood must not, however, reach the foot-rests.

The inclined edges on the holes for the foot-rests will best be made on the spindle moulder, though here you may need a small diameter spindle.

The girdle mould

Girdle moulds are easily made of strips of 1mm sheet metal. Sheet metal may, however, not be available in sufficiently long pieces and may therefore need to be joined with a soldered joint (see illustration).

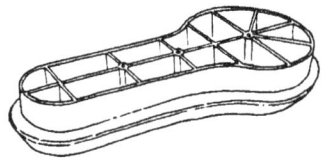

Plastic drop-hole mould

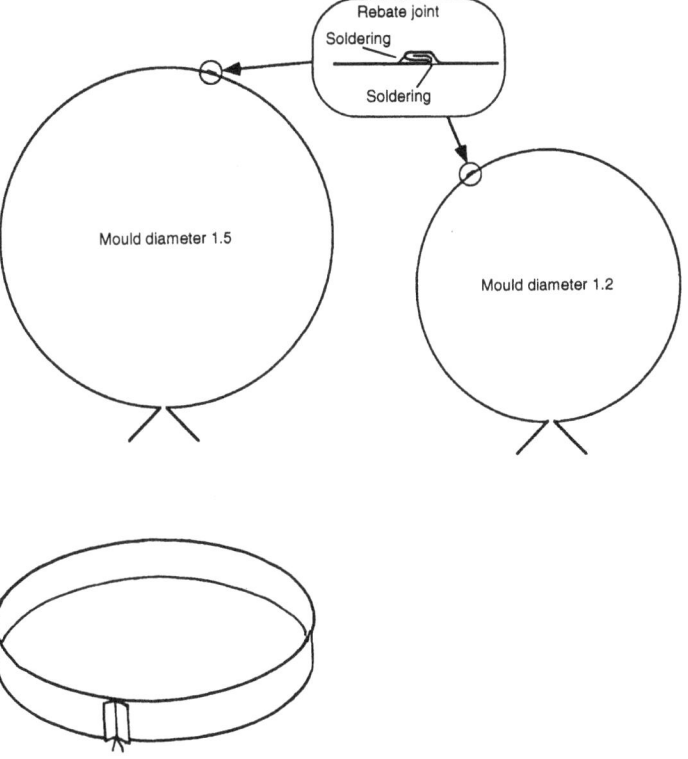

Girdle moulds (material: sheet metal, preferably 1mm; dimensions in millimetres)

Mould diameter 1.5
Cutting: Width 150mm, length 4910mm
Bending: 100mm

Mould diameter 1.2
Cutting: Width 150mm, length 3970mm
Bending: 100mm

Rebate joints
Sheet metal is not always available in the desired dimensions. It may therefore be necessary to join two pieces together with a rebate joint. As the durability of the moulds depends on the strength of the joints they should be soldered as the picture shows.

Tying of ends
After bending 100mm of the metal at both ends, the ends can be tied together with a soft string (e.g. sisal cord). Do not use iron wire as some elasticity may be required when 'lifting' the mould for casting the next one.

7. Slab casting with integrated SanPlats

This chapter explains step by step how to make slabs with integrated SanPlats.

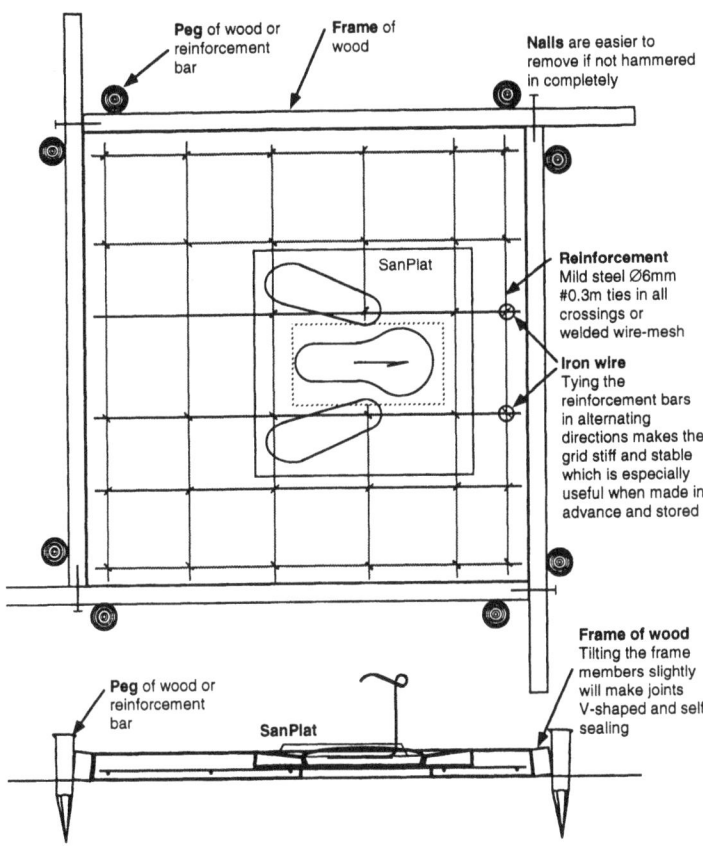

Casting a concrete slab with an integrated SanPlat

Casting of latrine slabs *in situ*

When precast slabs are not available they can be cast on the ground close to the pit.

1. Start by levelling the ground over an area slightly larger than the slab. Make a frame of planks with the

same width as the slab and the correct dimensions. With a tape-measure or a string check that diagonals are equally long and fix the corners with pegs in the ground (or stones) outside the nail-heads. (Round slabs can be cast with a mould of bricks laid in a circle.)

2. Cover the bottom with paper from cement bags, plastic sheets, banana leaves or similar material. Place moulds for holes at the correct places. If a SanPlat is to be integrated, wait until reinforcement bars are in place as the SanPlats should rest on them.

3. Cut the reinforcement (6mm) and put the bars in place (0.3×0.3m). Try to avoid extra cutting for the holes. Fix with iron wire or strings in all crossings and put small stones one inch (25mm) under the bars. Welded wire-mesh can also be used.

4. Mix the concrete on a clean, flat surface in the proportions 1+2+4 (cement+sand+gravel) if nothing else has been recommended.

5. Fill up to one inch deep with concrete (up to the bars) and compact the first layer, pounding it with a piece of wood. Lift the bars slightly to make sure that the concrete fills well under the reinforcement.

6. Fill up to the full height of the frame and compact again with a piece of wood. Level off the surface with a plank resting on the edges of the frame. Remove or add concrete as appropriate.

7. Finish off the surface with a steel-float and make sure that the surface is even and smooth.

8. Remove the frame as soon as the concrete is hard and cover it with sand. Keep the sand (and the slab) wet for one week without moving the slab.

9. To be sure that the slab is strong enough, test-load it after one week with the full expected load. (After four weeks it has reached its full strength, which is about double the strength it has after one week. If you want to test it after four weeks you should test it with double the expected load, to be on the safe side.)

Note 1: Dishing the slab

If a prefabricated SanPlat is not integrated, the top surface should be made sloping (dished) towards the drop-hole.

Note 2: Top surface

If you intend to put on a top-screed of cement mortar,

the surface of the concrete should be left rough but clean.

Note 3: Reduction of cement content
In a casting yard for continuous production the reinforcement and cement content may be reduced after testing.

Note 4: Avoid heavy slabs
Three-inch (75mm) slabs over 2.5m² (1.6×1.6m or diameter 1.8m) are difficult to move even very short distances.

Alternative ways of casting concrete slabs with integrated SanPlats

The simplest way to make a bigger slab with integrated SanPlat features is to use a ready-made SanPlat 60×60cm and cast the concrete slab around it. You can also use the SanPlat moulds when casting SanPlats with other dimensions.

1. Flat surface with a prefabricated SanPlat

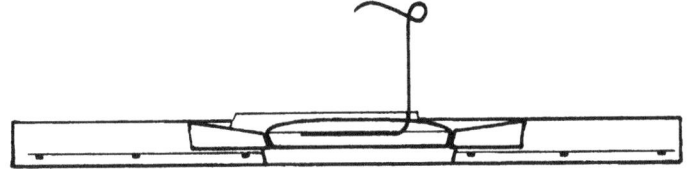

The slab is cast as a normal reinforced concrete slab. When arranging the reinforcement leave a space for the drop-hole. Pour the first layer of concrete so that it just covers the reinforcement bars. Put the SanPlat in place supported by the reinforcement, which should now be covered with concrete. Finish the casting as for a normal slab but with special attention to the fitting to the SanPlat.

2. Dished surface with a prefabricated SanPlat

The slab is cast the same way as before but the SanPlat should be placed on the ground from the start. The bottom edges of the SanPlat are chopped off for secure fixing.

3. Cast *in situ* with SanPlat moulds

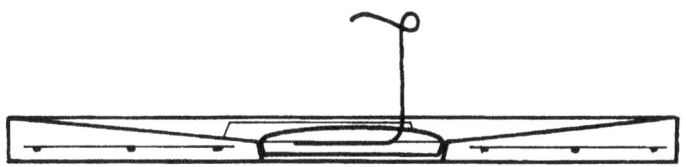

The slab is cast the same way as a SanPlat 60×60cm but with a bigger frame.

i

8. Handcarts

The hand-pushed cart made of scrap metal described in this chapter was extensively used in the Mozambican urban SanPlat programme where large, dome-shaped SanPlats were used. It was the introduction of transport services to the programme that made sales of dome-shaped SanPlats take off.

Construction of a 'tchova'

In the southern part of Mozambique (where Ronga and Changana are spoken) this type of cart is generally known by the name of 'tchova xita duma', which means, 'push...it will start'. Therefore it was decided to call the cart a 'tchova' (push).

These carts are made of scrap material and they are used for short distance transport of goods.

Our 'tchova' has dimensions adapted to the transport of slabs and other materials necessary to our casting yards.

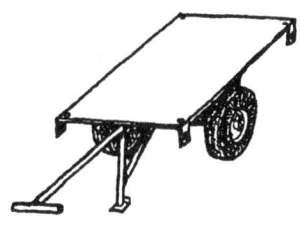

Pushcart for transport of slabs, known as a 'tchova'

Structure
The structure of a *tchova* is made of second-hand water pipes with external diameters of 32mm (1 inch) and 47mm (1½inch). It is possible to use other materials with the same resistance.

Platform
The platform is covered with 1mm thick iron sheet, which is folded and welded around the pipes. Note also the reinforced holes in the corners for fixing planks or poles to increase the load.

Wheels

The wheels should be the same size as for normal cars or small vans. Larger wheels will increase the *tchova's* weight, while smaller ones will be difficult to use on loose sand. The recommended diameter is 60cm. Before assembling the wheels, the brake pads must be removed from the wheel-house. You must also check that the wheel bearings are lubricated.

The wheel is fixed to the shaft with four flat bars (4×25×150mm) welded to the wheel-house and to the shaft, as shown in the diagram.

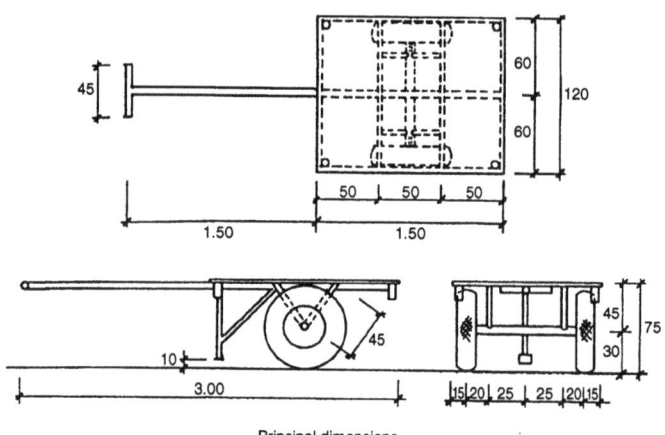

Principal dimensions

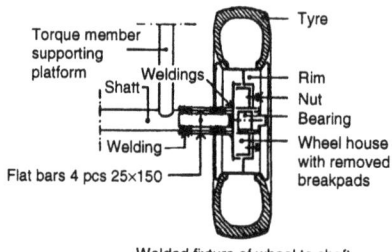

Welded fixture of wheel to shaft

Construction details for 'tchova'

9. How to implement a latrine building programme

A sound technology is the basis for the successful implementation of any latrine building programme. Successful implementation will, however, also depend on a number of other factors.

This chapter goes step by step through the most important aspects of setting up a programme:

- programme preparation
- programme promotion
- financial management.

Planning, monitoring and evaluation are covered in Chapter 12.

The preparation phase

Field programme managers
The term 'latrine building programme' does not necessarily mean a national building programme. It could also be a district programme or a village programme.

The principles are the same and should, therefore, also be known and understood by the field staff members. Field staff are programme managers and should be trained to be responsible for their part of the programme.

Know your target group
Normally programme managers have a fairly well-defined target group for their programmes. The better they know this group, the easier it will be to implement the programme. A few questions may help to focus on some of the most important aspects of the target group.

- How many families have latrines?
- Is there any problem with the existing latrines?
- Are they used? By whom? Always?

- How many families have no latrines?
  Why don't they have latrines?
- What kind of latrines do people have today?
  Are there different types?
  Why are there different types?
- How are they built?
- Who is building them?
- Who decides about latrine building?
- Who raises the question, and who has the final say?
- What do people feel about latrines?
- How much money, and how much of the families' own time, is used?

Questions like these should be included in a baseline study, i.e. a study assessing the situation before the programme starts.

Start by experimenting

Before you embark on a more systematic study, it is good to start building and experimenting with your ideas about the technology and methods of implementation. Keep your mind open for new and simpler solutions. Discuss alternatives with people you have confidence in. After this you will begin to know where the problems are and how to get around them. But equally importantly, you will have a clearer idea of what you do not know and what you need to know.

Make the baseline study

When the first latrine building programme is under way you may start your baseline study. The study has three principal stages:

- Preliminary study
- Data collection
- Data processing[1]

---

1 Data collection: In this case, interviewing households and filling in forms.
Data processing: Summarizing the collected information. It often happens that huge volumes of information are collected and that the project has no capacity to summarize all the information. Excellent computer programs can handle large volumes of information and find out relations between different factors. A data processing expert should in that case be consulted before the survey form is finalized and the full-scale survey is carried out. The person responsible for the processing may also have opinions about the survey forms.

The object of the preliminary study is to test your data collection form and to train your data collectors.

Try to use your own people to make the survey and, based on what you have discovered during your start-up period, discuss with them what kind of information you need and why. Make it clear to them that the object of the study is to collect the kind of information they need. An example of a form for the study can be found in the Appendix.

Develop your own form to make sure that you ask the kind of questions your people need for their own planning and the follow-up of their own progress.

The preliminary study should also test the data processing. In a small study you can count the answers manually. In a larger study computer assistance may be helpful. If you want to process your data, let a computer expert participate in making the form.

Verify the technology

Unfortunately we have too much confidence in what is written in books, and we have a tendency to be over-ambitious when we plan what to do.

Too many books (and consultants) have proposed building this, or that, type of latrine without having any idea of what people are actually building, what they can afford, what they are prepared to pay for, and what problems people feel they have with their latrines. Allowing people to see and comment on what you are doing may be a good start, but do remember that praising your demonstration latrines is not at all the same thing as paying for one. Most sanitation programmes are slow at the start, and many of them continue to be slow. If the programme goes too slowly it may be because your technology is too complicated or too expensive.

Learn from people's ability to build

To find out what people may be able to build, and like to build, take a close look at their houses. People rarely like to spend more time and money on the latrine than on the house.

You may find that the latrines use similar materials and building methods but are inferior in size, materials and workmanship. The reason for this is often that a latrine is considered to be a temporary structure which has to be replaced the day that it is full. Another reason may be that latrines have been built because of the insistence of health inspectors or other authorities, and not because people actually want them. If we are very successful, our latrines will look as attractive as the houses. If we try to make them more attractive than the houses we have gone too far.

Subsidies and the willingness to pay
Finding money may be a problem in rural areas. Many people may prefer to do the work themselves rather than pay someone else. In urban low-income areas, poverty is a problem and it may be wise to find out how much money has been spent on the existing latrine.

It may also be useful to ask people how much they are prepared to pay for a latrine slab, for a ventpipe, etc. However, you will only get reliable answers when people actually start buying.

The willingness of people to pay will increase as they understand and appreciate the new technology. Subsidies may therefore need to be bigger at the beginning of the programme and gradually phased out.

When there are no more subsidies, the time is right to release the whole programme to free enterprise.

Programme promotion

Advocacy[2] and political priority
Many latrine programmes are working with very little support. If latrine building programmes are to be successful they must be given capable people with good resources.

Reaching the key personalities

Working in isolation is another problem. Latrine building is not normally associated with high social status. Health people prefer to work in the clinic and water people prefer to work with pipes and pumps. A way to break this attitude is through advocacy; that is, convincing local influential people of the importance of the programme.

Advocacy should be carried out at top levels for top-level support to the programme, but it also has to be car-

---

2 Advocacy consists of the organization of information into arguments to be communicated through various interpersonal and media channels with a view to gaining political and social leadership acceptance and preparing a society for a particular development programme (McKee, Unicef, Bangladesh).

ried out at all levels. When the political will is there, the resources and the right people will come. If advocacy is ignored, or forgotten, the whole programme can quickly become a failure. Advocacy will, however, only be successful if it is based on a convincing and well-tested technology.

Social mobilization

Our own resources will never be enough. If the advocacy programme has been successful, it will be easy to establish links with other institutions and programmes to mobilize the whole community.

Schools may become engaged in teaching the importance of hygiene and good latrines. Older pupils can make door-to-door home visits, mobilizing families for latrine construction, while younger pupils can make latrine counts. Domestic Science teachers will most certainly teach personal hygiene. With assistance from the project latrine building can be made part of the curriculum. Biology teachers can explain about germs, disease transmission and the importance of using latrines. Mathematics teachers may organize monthly latrine counts and use the results in their teaching. Examples of counts of families with and without latrines can be made in absolute numbers and percentages.

Clinics and health centres may participate in the programme by educating mothers about the relationship between hygiene and health, and the importance of clean latrines for the health of their families.

Health inspectors and their field staff may be in charge of the latrine building programme, or support it by giving health talks and inspecting people's latrines.

Religious institutions may not only shown concern for people's souls but also act as Samaritans by helping people practically. Many pastors have used the pulpit to give health talks exhorting their people to build latrines. Muslims have very strict rules about personal hygiene and where not to defecate.

The key people may be the political leaders, and this brings us back to the advocacy programme and political priorities.

Increased implementation capacity

At the beginning, the staff allocated to the programme may be sufficient for the task. With massive promotional

Creating promotional capacity

support from various institutions the demand for better latrines will eventually exceed their capacity. Making priorities will increase your implementation capacity, as explained below.

Initially the project may offer to build demonstration latrines at very low cost. People will soon discover that this is a good way to get a good latrine at a very good price. As soon as you have a queue of clients it is time to give priority to people willing to do part of the work themselves. You may have started by building complete demonstration latrines with walls and roofs. Now you may give priority to people who are willing to build the superstructure[3] themselves, while the programme staff may build the pit and the slab (the substructure).

After some time you may have a queue of people asking to have the complete substructure professionally built. You may then limit your services to pit-lining and slab-making.

Eventually the programme may be providing slabs only. If the demand continues it may be time to raise the price of the slabs to a level where private contractors can take over production.

With increased demand, responsibility for latrine manufacture can be left more and more in the hands of the families themselves

A revolving fund[4]

The demand for latrines may be bigger than planned. The best way to meet this problem is by establishing a revolving fund, where revenues from the sales can be deposited. This money can then be used for unforeseen expenses and is especially valuable if the demand for SanPlats or other services exceeds what has been initially planned and budgeted for.

The revolving fund should be established at the beginning of the programme before it is really required, and you will discover that it is an excellent resource to meet unforeseen needs.

The revolving fund has saved many projects from disaster when central funding has temporarily failed

Privatization

The method described above is one of a number of pos-

Subsidies and privatization have to be considered carefully

---

3 Superstructure: Part of a building that is above the ground, such as the walls and the roof with doors and other fittings.
  Substructure: Part of a building below the superstructure: in this case, the pit with its lining and slab, etc.
4 Revolving fund: Normally a special account, often in a bank, where the revenues from the sales are deposited. Money from the revolving fund can be used to meet unforeseen expenses, but should in principle be used as a buffer. In that case the revenues will need to be used for buying more material etc.

sible strategies for privatization. Another strategy is to train contractors from the beginning and give them certificates of training.

A third method is to use trained contractors for making slabs and other latrine elements, and let the project sell them to the population, to begin with at strongly subsidized rates, and later at gradually increasing prices until the contractors can start doing business directly with the households.

In any case, the medium to long-term objective is either to make the project self-financing, or to let the private sector gradually take over.

Handling money

One good reason for privatizing is to ease the problem of handling money. However, this cannot be achieved overnight. When latrine building is part of a government programme, payment can be very complicated. Exceptional rules may be required to enable people to buy latrines or latrine components in a straightforward way from the project. This sort of control is nevertheless a key issue as thefts may be demoralizing for the whole programme, and programme officers may have to spend too much time investigating irregularities.

Watch out for creative bookkeeping!

Budgeting

All sanitation projects need some form of funding. Revenues from sales are rarely enough to cover all expenses. Part of the expenses is often covered by government or other organizations and these institutions have fixed dates to receive applications for the next year's funding. Missing one of these dates may imply a year without funding or with a shortfall in funding.

Even here the revolving fund may be helpful. If the budget goes wrong you may be able to continue programme implementation using the revolving fund.

Do not miss the date for handing in your budget request

Planning and keeping records

For your own preparation of the next year's budget you need a plan for your production and your expenditure. Note that budgeting is normally carried out half a year before the fiscal year comes to an end. In many cases it can be difficult to foresee what will happen next year. A good approach is to predict the minimum income that could be expected from the project.

Planning for the future is easier if you have clear records of what you have already done

This guess should be based on the present year's production. You would need to have at least monthly records of what you are producing and spending to enable yourself or anyone else to make a forecast for the coming year's production.

Making a graph

Making a graph or a chart on squared paper of your monthly sales is probably the simplest way to see how sales are going. You should probably also make another graph on the same piece of paper to show what your targets are and how you will eventually serve the whole community.

Graphs are a good way of making figures and trends accessible

10. Promotion and hygiene education

The principal reason for building latrines is to improve health. For the individual family, reasons such as status and convenience may be equally or even more important. Effective promotion of improved latrines draws not only on health arguments, but on all the practical and emotional reasons people may have for building and using a good latrine.

This chapter explains how to promote construction of better latrines:

- by using arguments that people understand and which fill felt needs;
- by winning the interest and commitment of the influential EER-group (the Educated, the Employed and the relatively Rich);
- by educating your own staff on the importance of their job;
- through hygiene education.

It also outlines a possible promotional strategy and goes through the procedure step by step.

Children's faeces are up to twenty times more dangerous than those of adults

Convincing people

If you have followed the guidelines in the previous chapter you should now be well prepared for expanding your latrine building programme. The objective is to convince each and every family to build or improve their latrines, and this can only be achieved through public awareness and a generally felt need for improved sanitation.

This need is created through *promotion*, by emphasizing arguments like privacy, convenience and being modern. Such arguments are, however, not enough if we want to create general awareness and put peer pressure

There are many other reasons besides health that can be used in a promotional programme

on resistant families. Eventually it is through *hygiene education* for the whole community that 100 per cent coverage will be achieved.

What is the difference?
Promotion and hygiene education often mean the same thing. People should be convinced about latrine construction through an *understanding* of their health benefits. Promotion and hygiene education do have different objectives and their success should be evaluated in different ways. The success of a promotional programme is measured in the *number of latrines built*, while a hygiene education programme is evaluated in what people know and how their *attitudes and behaviour have changed*.

Promotion changes attitudes

Hygiene education changes behaviour when attitudes have been changed

Promotion has a short-term objective, while hygiene education has a long-term objective. They support each other in working for better hygiene and health. The boundaries between them may also overlap.

Being modern
Promotion uses emotional messages to a great extent, aiming at changing people's attitudes in a subconscious way. Many examples could be taken from the commercialization of consumer products, like soaps and hair shampoos. We can also look at sales promotions for cars and radios, hi-fi equipment and similar products, where our own status may be reflected in what we buy, and where, if we do not buy, we may indirectly look inferior or old-fashioned.

Emotional messages are often more effective than logical ones

Most people want to be modern. Being modern is linked to development and high status. This is a strong selling argument. A SanPlat should therefore look modern and be marketed as a modern product. A slogan like: 'A modern latrine has a SanPlat' says indirectly that latrines without a SanPlat are not modern and the people using them are 'out'. The positive way of putting it is: *A modern person has a latrine with a SanPlat*.

Status, privacy and convenience
The relation between hygiene and health is often complicated, and people have experience of their children still becoming ill even if they have a latrine. Reasons such as status, privacy and convenience may be easier to understand for many people.

Most people use latrines without even thinking of the health aspect; it is just a good habit

The privacy aspect is important for all people. Promoting the use of a proper latrine for privacy reasons may be a good selling argument. Few men want to see their women defecating publicly, and many women have to suffer by leaving the home in the darkness of the night or very early in the morning to relieve themselves.

Promotion can also use messages about pleasing other people: 'Your visitors will love it'. Offering a dirty, poorly built latrine to visitors may be embarrassing. A SanPlat in your latrine may help you out of an embarrassing situation. Many parents insist that a young man must have a house with a proper latrine before he may be allowed to marry their daughter. 'Even your mother-in-law will love your SanPlat latrine.'

It *is* convenient to have a latrine and it actually does provide privacy, though this may be interpreted as a luxury by some people.

The SanPlat Latrine: Your visitors will love it! No smell, no flies, no embarrassment

Hygiene and health

The health arguments may be more important but they are more difficult to sell. We all know that not all diseases are controlled by hygiene and health education, and that even if this were the case such changes would take considerable time. Still, the argument that good sanitation improves health is very valid and must be emphasized if we want to achieve improvements in public health in our communities.

Attitudes and the spoken word

The spoken word, person to person, has proven to be the most efficient method of promotion and hygiene education. If you have been successful, people will have good things to say about your latrines, and the message will pass from one person to another. Eventually everybody will know. How much they will know depends very much on how clear you have been in the explanation of the programme. Even more important is what people feel, and that depends on how you present the message and how you relate to other people. That is what makes a good promoter and health educator – a person who has a winning personality, though that can be learnt.

Knowledge, attitudes and practice (KAP)

Many readers will have heard about KAP studies, but KAP is also a process in behavioural change. KAP stands for Knowledge, Attitudes and Practice. In changing our behaviour we always go through the same process.

First we learn about something new (in this case, new and better latrines). We now have the *knowledge*.

The next step is that we learn to appreciate the new latrines: we change our *attitude*. We would like to have one, but... There may be a number of reasons why not yet. It may be money, it may be not wanting to be the first one to start something new, or there may be other reasons.

The final step is *practice*, when we take the final decision, buy our SanPlat and install it in our old latrine or build a new one.

The KAP process takes time, and you should not be discouraged if it takes some time before you see the practical impact of your work. People may still be learning and appreciating, waiting for the final push to come – to go from the K and A level to P.

Winning the EER-group

In a rural community, the most influential people are the Educated, the Employed and the relatively Rich (the EER-group). To become successful promoters, you need to spread the knowledge as well as possible throughout the community. You want people to appreciate the new latrines and to buy and install SanPlats.

The best way to achieve this is to win the *active* participation of the influential people, the EER-group. These people can afford, and are often interested in testing, modern solutions. They are happy to influence other people. They are literate but also critical. Win them and you will win the whole community.

The example shown by others gives a very convincing message

Media and messages

Simple and straightforward messages are important. Posters and pamphlets can be powerful if correctly designed and distributed. Local testing of pictures and messages to support your promotional strategy is important so that you are sure positive information is reaching the people.

It is obvious that the EER-group will and should be the first to understand your poster, to read your pamphlets and, probably, also to buy SanPlats. The EER-group have radios and they can afford to buy batteries to listen to them.

Radio programmes do not reach everybody, and TV programmes reach even fewer, but publicizing the latrine building programme on radio and TV gives it a status and a legality which will support whatever is presented person to person.

The majority of people may not have money to buy books about hygiene and health. However, some *will* have the money, and the fact that such books are paid for gives them status – they will be kept in a safe place and read a great many times.

Radio and TV programmes may only reach a few, but will lead to discussion

Enforcement

In an urban low-income community the population tends to be relatively well educated. People know about hygiene and health. They know about latrines and the majority have them. However, people in urban areas tend to be less responsive to peer pressure than those in rural communities.

Poor hygiene is more dangerous in high-density urban areas. Legal enforcement of latrine building will therefore be justified and understood. For enforcement to become meaningful there must be a general understanding that latrines are important for people's health. Hygiene education must therefore be given high and urgent priority.

Enforcement is, however, only effective if the majority of the population already understands, appreciates and practises the rules we want to enforce. If not, there may be a conflict between what the programme wants and what the people want, and such a conflict may become very difficult.

Enforcement is a good tool only when most people already have and use latrines; it is very effective in peri-urban areas

Saving more lives than the doctors

Understanding why people get sick, and why latrine building and hygiene are so important, may be a very strong motivating factor, *especially for the staff involved in the programme*. If your staff members understand that their job will actually save many people's lives, they will be better motivated, and with better motivation they will do a better job.

Working with latrines is often regarded as an inferior task, but as a matter of fact, people involved in building latrines save more lives than doctors. If you make your staff members and the community understand this, they will feel proud of their jobs and perform miracles.

Hygiene education

Changing people's habits in a whole community is a task which takes a long time, and requires special qualifications which go beyond the scope of this book. However, that does not mean that we cannot do it. On the contrary, it is through understanding (*knowledge*) that we will change people's *attitudes* and behaviour (*practice*). Below, you will find a very short lesson in hygiene education.

Mothers are the best hygiene educators in the world

Many people suffer from disease

Diarrhoeas and parasitic infections are, together with respiratory diseases, the major killers in developing countries (they kill more people than malaria).

A person who is weakened by parasites will more easily contract other diseases, will suffer more and is more likely to die. Parasites make us feel tired, depressed and unhappy. Diarrhoeas and parasitic infections can be reduced if we build, and use, latrines properly.

The vicious circle of disease, malnutrition, poverty and economic stagnation must be broken

How a disease develops into an epidemic

When many people get the same disease this is called an epidemic. For example, if faecal matter is lying around, it will be washed into streams by the rain. Animals will also spread it to places where children play, and they will get sick. Mothers treating their sick children may get their hands contaminated by faecal matter, which will then pass to the food, and the whole family may be affected. Having no latrine the family will spread more infected faecal matter into the surroundings, making more people sick, they will cause still more contamination and so an epidemic is established.

Disease creates more disease

Protection of water sources

Water in rivers and ponds is frequently used for bathing and cleaning. Very often the same rivers and ponds are also the supply of drinking water for many families. If latrines are not used, fresh faecal matter will be washed into streams and ponds by heavy rains. Epidemics of cholera and infant diarrhoeas are very often caused by people not using latrines, but normal diarrhoeas and parasitic infections can also be spread this way. Any latrine which is used is therefore very much better than open defecation.

Polluted water carries disease

Animals spread disease

Many animals eat human faeces: pigs, goats, chickens, dogs, etc. They may seem to be cleaning up the environment, but as a matter of fact they do the opposite. Bacteria and parasites pass through the bodies of the animals, which will spread them to all the places where they leave their droppings, such as in wells and gardens.

Building better latrines

By building latrines we prevent faecal matter from reaching the water courses, and chickens and animals will not be able to spread disease to places where children play. Better latrines are easier to keep clean and people appreciate them more.

Use of latrines prevents disease

Washing hands

Building and using latrines is important, but leaving the latrine with dirty hands will transmit diseases directly. Therefore, washing of hands before touching food, and *always* after having been to the latrine, is very important. Dirty fingers are very dangerous!!

Dirty hands are very dangerous

Other diseases

Latrines are necessary. Unless we use latrines we cannot prevent diarrhoeas and parasitic infections from spreading. There are, however, other methods of disease transmission. For example, some diseases are spread from person to person directly. Some diseases are not related to personal hygiene. Therefore, we cannot expect all diseases to disappear, but we can bring about considerable improvements.

Promotion

Winning women to win the men

The decision to build or not build is often taken by men, although recently more and more women have engaged themselves in building latrines. For more men to decide to buy a SanPlat and to install it, or to build a completely new latrine, they need persuasion. They need to know, they need to appreciate, and then they need the final push to do it.

It may be easier to convince women of the need for latrines, but building matters have traditionally been in the hands of men. Winning over the women may be a

way to persuade the men, especially if other influential men have already taken the same decision.

Mothers are caretakers
Mothers are concerned with issues of health, hygiene and cleanliness. It is also mothers who are normally the main educators of children. Women learn rapidly and appreciate new and better things for the house. Wives will influence their husbands in the question of building or improving latrines. They need and deserve all the support we can give them.

Women are the key to involving men and children... and more women

School children as 'agents of change'
The school plays an important role in the development of a village or town. School children learn things in school which their parents never learnt. They are eager to learn and have a high status in their families, especially if they have more education than their parents. They are the real agents of change in society. School children are therefore an important target group for promotion and health education. The most effective way to reach school children is through their teachers.

Children teaching children
In most families older children take care of the younger ones in the family. It is to a large extent older children who teach and train their younger sisters and brothers. Since children are health educators, they deserve to be recognized as such.

Children are teachers of children and should be treated as very influential people

Elders may have a final say
In many communities the elders always have the final say. The elders may be conservative and sceptical of modern things, but ignoring them may have disastrous effects on the whole programme. Often it may be wise to visit the elders first to have their approval of the programme before you go ahead.

Everybody is important
All ages and all sexes play an important role in promotion and hygiene education. We will achieve the best impact if we take care of each group in a way which corresponds to their role and ability.

See everybody as a trainer
Everyone, regardless of age or sex, participates in

training others. Even the children are trainers. Explaining their important role as trainers will have a positive impact. They feel respected and important, and will do their best to live up to this important role. Mothers appreciate receiving teaching skills. They will love to be looked upon as colleagues. Children will grow to adults in their role as a trainer, and will try to behave as an adult, providing a good example to sisters and brothers.

And yourself

It is extremely important that you have a positive attitude to yourself, to the people you are working with, and to the job you are doing. The right attitude can be formulated in three short phrases:

> *I'm good*
> *You are good*
> *We can do it!*

Your own attitude is clearly important

Looking at yourself in a positive way is the first and most important step in person-to-person promotion. Looking at others in the same positive way is the next step. If you manage to influence your staff members and your counterparts to support each other you will be able to achieve miracles together.

If you want to learn more about how to work with people you can turn to Chapter 11.

A promotional strategy

There are principally three phases in the implementation of a sanitation programme:

Phase one

Phase one is targeting *the influential group* (the EER-group) with the object of obtaining allies at local level for implementation of the programme.

Considerable work should be invested in this group. They know a lot about improved latrines (Knowledge) and appreciate the need for environmental sanitation (Attitude). They should all practise their acquired knowledge using the improved latrines (Practice).

Once the influential people have and use improved latrines, a positive image is created. They are modern.

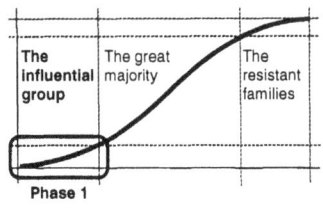

Phase 1

Phase two

The object of phase two is to convince *the great majority*

of the population. They will be convinced by the influential group, based on messages like privacy, convenience and being modern. Health education will have taught them that improved sanitation is an absolute need for the whole community.

Phase three

The object of phase three is to convince *the resistant families*. Through hygiene education, peer pressure will have been put on reluctant families who consider the use of latrines a luxury. They eventually have to accept the idea that everybody should have and use an improved latrine. Enforcement is now possible, if necessary.

Knowledge, attitudes and practice

Each of the three groups will have to go through the three stages of KAP:

- K   To gain Knowledge of the advantages and to understand the need.
- A   To gain a positive Attitude towards the required improvements and understand the eventual need for everybody to use them.
- P   To adopt improved sanitation Practices, which includes the construction of improved latrines and improvement of existing ones.

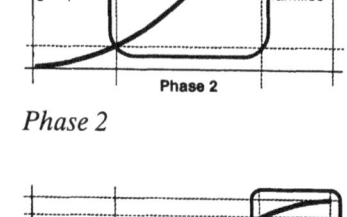

Phase 2

Phase 3

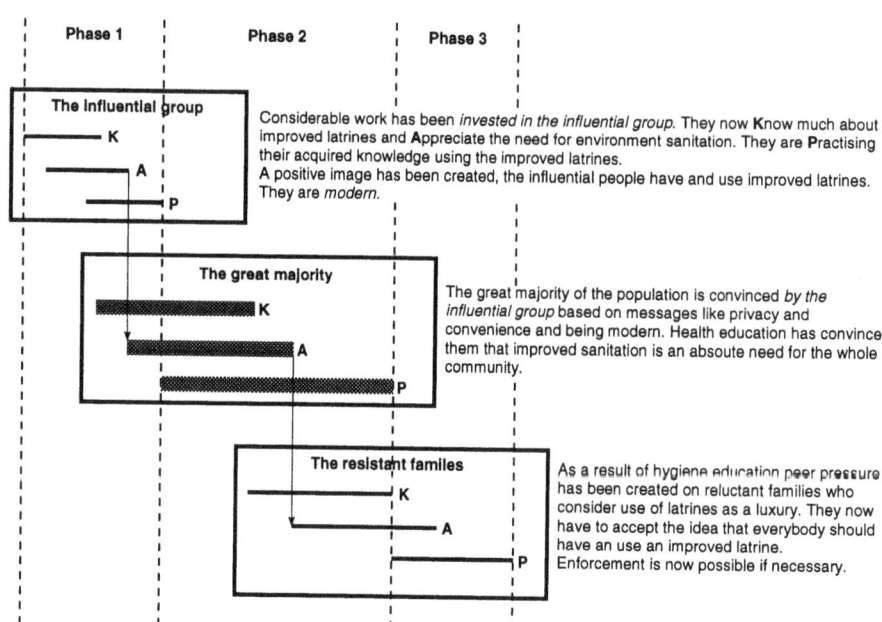

The phasing of the implementation process

Timing
How long the different phases will be depends on the community and the resources allocated for promotional activities. It is, however, crucial that the project is able to deliver at the same pace as the community is responding to promotional efforts. (See also page 83: increased implementation capacity.)

Step by step
Let us assume that you are a project officer with the task of introducing improved latrines in a certain area.

Listen, see and understand
Start by finding out the needs: the real needs – that is, the needs people actually have – and the felt needs – that is, what people *feel* they need. This can best be done by talking to people with good contacts in the communities *and* through seeing the reality. Try to understand what people feel and what it may be possible to do.

Make contacts
Before any major practical steps are taken, establish a network of contacts with influential people who you believe may play an important role in the implementation of the programme. Keep a low profile to start with as there is a risk of disappointing people if they imagine more than you can promise.

Advocacy at local level necessary

Select a pilot area
To get a good start it is important to start in an area where you are sure to have success. Avoid all problem areas at the beginning, including social problems as well as technical problems. Explain the importance of starting in an area where other problems do not jeopardize your efforts.

Start where you are sure to have success

Establish the delivery system
Get established. Make sure that you have something real to show. People want to see what you are talking about and, if it is to be produced locally, how it is being made. Remember that your casting yard and your materials depot will become your showroom. They should look neat and tidy and well organized. The siting is important; people should see it. Secure your supply lines. Prompt

Do not make promises unless you can deliver

delivery is a must. Remember that all excuses are bad excuses, even if delays and faults are caused by others.

Win the key people
The first step is to win the influential people, but *with the object of serving the whole community*. There is normally a very small group of two or three people who are the practical leaders of a society. They may be the party chairman, the administrator, the church leader or the traditional chief, or there may be others. Talk to people to find out. You should also find out the best way to be introduced to them.

Cultivate your contacts with key people

Discuss strategies
Discuss possible strategies for achieving 100 per cent coverage.

Discuss technical solutions and upgrading possibilities. Discuss the general health status in the area and the danger of people not using latrines. Try to identify the problems as early as possible and discuss alternative solutions.

Discuss the possibilities of engaging schools, churches, women's organizations, village health committees, the party, health workers, shopkeepers, etc. Do what you can to assist them to get their latrines built or improved, but remember that public and institutional latrines need different designs from private ones. The object is to provide good examples, not to show off.

Is there anybody in the group of influential people who is against the programme? If so, why? Older people can be both conservative and influential. Maybe they do not like the idea of sitting in a house to defecate. Perhaps there should be unroofed latrines. Do whatever you can to get the conservative people involved. If not, you may run into serious problems as they may discourage others from supporting the programme.

If influential people participate in strategy discussions, they will help you to implement them as well

Win the EER-group
There is always a group of people who play an important role in the community. In the rural areas they are the Educated, the Employed and the relatively Rich (the EER-group).

Make your assessment of their opinions, knowledge, attitudes and their own sanitation practices. Give them promotional material like posters, pamphlets, and

Listen and judge; be selective in what you believe; double-check all information

hygiene education materials. They are the people who can and will read. Possibly they could study parts of this book. The more they know, the better the response you will get.

Warning 1
Proposals are often made to build latrines free of charge for especially needy families. This is not to be recommended, especially at the beginning, as needy families normally have a low status. Starting with the needy families may give a low status to the new latrines, hence jeopardizing the mobilization process. The needy families may receive subsidized latrines once the programme has already gained momentum.

There is a risk of latrines being associated with poverty

Warning 2
Do not build demonstration latrines free of charge for anyone. This will always cause problems as other people may feel it unfair and stop participating. Others may just wait, hoping to get a demonstration latrine free of charge as well.

Progress may be slowed down by building demonstration latrines free of charge

Continue with the great majority
If you have your delivery system established and the EER-group mobilized, it is time to mobilize the community organizations, hold meetings and discuss who is doing what and when. Arrange further meetings to follow up: what has been done and what is missing? Are there any problems? If so, why and how can they be overcome?

Hygiene education
Practical hygiene education takes a long time, but is absolutely necessary if people are going to understand why it is important that everyone has and uses latrines. The latrine does not have to be beautiful. It is the proper use which is important. The general appearance of the latrine may, however influence people's attitudes, and attitudes will influence behaviour.

Hygiene education takes time

Hygiene education can be carried out through women's organizations, schools, churches and mosques. Health personnel would, of course, play a key role in educating everybody.

Peer pressure
The community is now well aware of the dangers of

The community will put pressure on reluctant families

human faeces and the need for everybody to use latrines. People who up to now have considered latrines a luxury will feel and possibly also understand that they must build and use latrines.

Special cases

There will always be families which for different reasons are not able to build a latrine. These may be widowed or sick people, single mothers, etc. Given that the community now understands that it is in everyone's interest that even the remaining families use latrines, the community may feel it has a responsibility to assist.

Some need help, others need more pressure... from the authorities

11. How to work with people

To work with people is a skill which can be learned like any other; the fact that some people seem to have the skill naturally means that teaching it is often forgotten

Management

The implementation of low-cost sanitation programmes may involve management of the casting yard where SanPlats are produced or it may involve many other functions such as latrine building, teaching people how to build improved latrines, promotion and hygiene education.

This chapter addresses itself to all levels of the large and important group of sanitation programme managers. It will probably also be of interest to other people in managerial positions.

Some of the advice might seem elementary, but it is included because most sanitation programme managers have never been in a managing position before. It is based on some of the latest material in the field of personnel management and may interest more experienced managers.

You – the manager

As manager, you have overall responsibility for the implementation of improved sanitation in the area where your programme is located. There is nothing that can be excluded from your responsibility.

Three things are, however, of special importance in your new role as manager:

- People   (Personnel management)
- Production (Getting the job done)
- Papers   (Keeping your records)

This chapter concentrates on the first element, People, which is your most important resource.

Personal personnel management

People

All professionals use tools in their job. The manager's principal tool is people. As a manager you have to learn to work with people. Some people have a natural talent for this – lucky them! Others have to learn it step by step, as they would learn any other skill.

The manager's principal tool is people

To learn to work with people it is necessary to start by looking at yourself to find out what sort of a person you want to be.

Winners and losers

Some people are more successful than others. They seem to have the ability to grab the chance when it appears. They are winners. Other people are permanent losers. Why? To a great extent it depends on your way of looking at yourself and the way you look at the people around you.

I am good

Supporting *yourself* is the first step. We all have two voices inside ourselves. One of them says:

- It's OK
- No problem
- I can do it.

I am good...You are good... so we can do it!

The other voice says:

- I don't like this
- It's difficult
- Somebody else should do this.

People who listen to the positive voice have a tendency to become winners simply by believing in themselves, while others might have a tendency to become losers.

Conclusion: Encourage yourself to become a positive person – a winner.

You are good
The next advice concerns your attitude towards *other people*. There are people who try to gain self-confidence by talking about other people in a negative way, as inferiors. Step by step they develop a negative attitude to people. This will affect their relationship with those around them.

People with a winner's attitude and good self-confidence can afford to see the positive sides of other people and support them.

+ I am good
+ You are good
= We can do it

This might seem to be a very rough simplification of the

relationship between people. Of course, the subject is more complicated than that, but we must start somewhere. Build self-confidence – your own and that of others.

If you do not agree, start looking for another job.

Four-minute relationship building

As a manager, your main task is to help people to do a better job. We have already mentioned the importance of supporting people's self-confidence. The following advice is based on European experience. Check for yourself how relevant it is in your own country.

It has been scientifically proven that when you meet a person the first four minutes are the most important. What you achieve during these first four minutes almost totally determines how the rest of your relationship develops.

So, what should you do during these four minutes?

1. *Smile*
   Smile to show that you have a positive attitude to the person you meet.
2. *Establish eye contact*
   Look at people when you talk to them, as long as you feel that this does not embarrass them.
3. *Use the person's name*
   We all like to hear our names. It makes us feel good. (I am good + you are good.) Try to find out the person's name and use it directly. By hearing yourself saying the name you will remember it more easily.
4. *Pay undivided attention*
   During these four minutes the person you are meeting is the only person that exists for you! What do they want from you? Are they happy? If not, why not? Remember that you have four minutes to win their confidence.
5. *Observe the person's posture and attention*
   Why? A person's posture can tell you a lot about how a person feels. Are they tense or relaxed? Are they interested, or are they thinking about something else? Reading body language will tell you much more than just listening.
6. *Show that you respect and accept the person*
   Address the person in the way that their tradition requires. If the person does not feel respected and

You must be a positive person

accepted they might feel bad, and that you are responsible for this – and that may grow into an obstacle between you.

7. *Show that you respect your colleagues and the organization you work for*
   As a manager you represent an organization. People are coming to see you as the representative. If you show disrespect people may lose confidence in the whole project, in what you are working for and possibly in you as well.
8. *Be calm and confident*
   Even if you feel worried and nervous, even if things have gone extremely wrong, people expect a manager to be calm and confident.
9. *Be groomed, clean and tidy*
   Make it a habit – it will help you in your work with the people around you!

Furthermore, it is much easier to build and maintain good relations with people if you are relaxed, have slept well, and are satisfied and happy with yourself. If you are not, try to find out why and to make changes!

Show that you respect and accept the person

Leadership

When a leader is good, people experience the feeling 'we did it ourselves'. They feel proud of their work. They feel good.

As a manager you will often discover that time is short. You will feel that you need more time to spend not only with your staff but also on planning and administration. To increase capacity you have to delegate tasks to your subordinates.

Development level

Delegation assumes that your staff have a high level of competence, which is not always the case. Another factor is commitment. Roughly we can divide our staff into five different groups of development.

Level	Development
D0	Low competence and low commitment
D1	Low competence with high commitment
D2	Some competence but low commitment
D3	High competence but low commitment
D4	High competence and high commitment

Each one of these categories will require a different style of leadership from you. You have to treat people according to what they know and according to their attitudes to their job.

People who do not yet know their job or a particular task will require direction. They need to be told how to do the job and to be closely supervised while doing it. On the other hand, a qualified staff member will become irritated by too many explanations and by being checked up on. A staff member may, however, be very qualified but uninterested in the job, or uninterested in a particular task. Different levels of commitment or motivation will therefore also require varying levels of supervision, but of a different type. It is important, but difficult, to identify reasons for low motivation. Whatever the reason might be, it can usually be compensated for by supportive, encouraging supervision.

We shall discuss more about different reasons for low motivation later in the chapter (page 109).

Leadership style

The different combinations of competence and commitment will require different styles of leadership.

Low competence and low commitment – D0
This is the problem case for any manager: not able and not interested. Any D0 staff members should be moved from their present tasks to something they are able to do. They then become D2, ready for a coaching leadership style.

Low competence and low commitment

Directing leadership – D1
For level D1, low competence and high commitment, a directing leadership style is required. A staff member who is highly committed to the task will be motivated to absorb all the guidance you can give and will rapidly develop competence.

Coaching leadership – D2
The D2 level is more difficult. Some competence combined with low motivation often creates a difficult attitude to work. Staff members in this category might think that supervision is not necessary as they 'already know their job' and will be irritated by somebody checking on them all the time.

However, people at this level may be willing to discuss

Low competence, high commitment

targets for their work. As they are not very competent the targets should not be put too high and the agreed time for inspecting should not be too long. In order to defend their prestige as staff who know their job, commitment will rise and they will most probably work very much harder.

Shorter intervals between inspections will give many opportunities for evaluation and constructive criticism. If a staff member's commitment to work is doubtful you must be very careful how you criticise. Give close attention to progress and give plenty of positive praise for this. Avoid negative criticism as poor competence often goes together with poor self-confidence.

Contracting leadership – D3
Competent but poorly motivated staff members (D3) have a similar profile to the above (D2). In this case, however, high competence must be respected. Staff who feel that even their competence is doubted are likely to become even less motivated.

'Contracting' is, in this case, a useful method. Contracting means discussing targets and setting goals. If your staff members are aware of their competence they will set their targets high and go for it. Your leadership can therefore be limited to listening and supporting.

If your contracting is successful your staff members might develop to level D4: competent and committed. They might however need some motivation to raise their commitment, in order not to drift down to D3 level again.

Delegating leadership – D4
Competent and committed staff members (D4) will require very little from you as a manager. Their work will advance without your interference and they will find stimulation and motivation in what they are achieving through their work.

With your competent and committed staff you will only need to discuss general guidelines. If you do not give guidance, they may request it themselves or even help you to elaborate guidelines.

Peak performer

There is a risk, however, in just leaving the D4 working alone. They may have gained their high motivation through your support. Leaving them alone might result in them sinking to D3 level.

Task-specific leadership

We have now discussed different types of leadership styles related to different levels of competence and commitment.

Some staff may, however, be good at and interested in one field whilst bad at and uncommitted in another field of their work. They may therefore be treated as a D4 for one job and as a D2 for another. The obvious advice is to try to give the right job to the right person. An incompetent builder might become a good promoter and a sleepy night guard might like to learn how to build.

A sleepy guard might like to learn to be a builder

Promotion of staff

It is common practice, and a good rule, to promote successful and reliable people, so-called 'peak performers'. Remember, however, that promotion leads to new tasks in which the person may be less competent and possibly less interested.

Good builders do not always become good foremen. They might become impatient with less skilled builders, who as a result might lose their interest and motivation in their work.

The best builder, earlier a D4, competent and committed, able to work completely unsupervised, may now become a D1 or D2 and will require directive or coaching leadership from you.

Summary

The key to successful leadership is:

- Do not have the same leadership style with all your staff members.
- Try to vary your leadership style with the same person.
- Your style of leadership should depend on the competence and commitment demonstrated for each different task.
- The style you choose should depend on the staff member's task-specific competence and commitment.
- Pay plenty of attention to assessing your staff members and...

Be personal!

Psychology

How people react is itself a complete science. We will only touch on the subject with a few points which are of special importance for your work as a manager.

Motivation

All of us have reasons for doing what we do. The reason is called the motive. One of the psychologists who has looked into people's motivation is Maslow.

Maslow sees our motives in a hierarchy, or a staircase of needs of different importance. Your basic motives are to satisfy needs such as hunger and thirst. Only when these more basic needs are satisfied will the person be motivated to satisfy needs of a higher rank, e.g. security, social contacts and self-esteem.

The family situation can both motivate and demotivate a person

Self-esteem through work (doing) should be the most sophisticated or highest form of motivation. We find it among the peak performers (D4). Imagine a good worker who suddenly has problems with his wife or his girlfriend. His motivation will immediately concentrate on this more basic need, rather than on his work. Consequently, basic problems, such as hunger or thirst, can cause drastic drops in the motivation and performance of a worker.

The corresponding advice is that you must try to be sensitive to what your staff have on their minds. It may be difficult to judge whether to interfere or not, especially if the matter is 'none of your concern'. A personal relationship with your staff, will, however, make the situation easier; they will feel you are a friend as well as a boss.

Strokes

Strokes are any kind of attention you may get from or give to another person. There are positive strokes and negative strokes. And there can be complete absence of strokes.

Positive strokes are good news, things you like to hear or receive, while negative strokes are bad things that you do not like. Worst of all are zero strokes – being ignored.

Strokes, especially the positive ones, are the most useful tools in your work as a manager. Pay attention to people. Even if you have no time to go into details in their tasks or problems, use positive strokes to show attention. Be careful with negative strokes. And finally: do not ignore anybody.

Do not ignore anybody

Pacing

Some people go well together. They like to do things the same way. Other people just do not fit together at all. In addition, days and situations can be different. The phenomenon is called 'pacing' and is to a large extent based on non-verbal communication. To pace a person is to follow him or her in body language, in speed, in rhythm, in thinking and in talking. To pace a person is a very effective way to build up a situation of confidence. In the same way, shouting, interrupting, not answering or not paying attention is an effective way to create irritation and bad working relations.

We have already talked about the initial four minutes and relationship building. Pacing should be the natural ingredient at every step. If you feel that the person you are dealing with is excited – follow them, be excited. If someone is sad or disappointed, try to keep the same speed, the same tone and pitch of your voice, body movements and position. They will appreciate it.

If the person you want to have a talk with is sitting – sit down. If they are standing – stand up, or ask him to sit down. Go through all the steps in the four minute relationship building, trying to pace the person as much as you can.

Try to make it a habit to pace people. They will not realize you are doing it, but they will appreciate it unconsciously.

Some people get on well together

Stress

Stress is a normal and necessary reaction of the body to something that may concern us. The factors that are stressing us are called stress factors and it is these that motivate us in everything we do. The body's reaction to stress is to increase production of adrenalin (a hormone produced in the kidneys) which, among other things, gives us shallower and quicker breathing, tenser muscles and a sharpening of all senses. The body is prepared for action.

These normal and necessary symptoms disappear as soon as relevant action is taken, or the reasons for our stress have disappeared. When we talk about stress, we usually mean abnormal and prolonged stress, caused by permanent stress factors. This kind of prolonged stress can and will cause negative effects, mentally as well as physically.

Stress, if caused by an external factor, will draw our attention from our work, we will be absent-minded and have difficulty concentrating. Prolonged and high levels of stress can also cause nervous problems, stomach disease (ulcers), alcoholism and reduce our resistance to other diseases.

The most effective way to reduce your stress is to identify the stress factors and eliminate as many of them as possible. Those you cannot eliminate you should try to evaluate.

Imagine for example that you have a conflict with a colleague who might be talking about you behind your back. First you may assess how serious the gossip is, what is being said about you, and how serious are the consequences it may have. Normally the consequences are less serious than you first thought. And even if they are serious, you are probably strong enough to cope with that also.

The next step may be to solve the problem with your colleague or decide just to ignore it. (Note: Do not ignore your *colleague*! Ignore the *problem*!) Maybe techniques mentioned in this manual, such as strokes and pacing, can help you to sort out the situation without any drastic measures. If you are a good and honest person your colleague will probably realize this and change his or her behaviour. Being able to forgive is a sign of a good leader.

Conflicts

There will always be situations where people do not agree and become angry.

Aggression is a very normal reaction to high levels of stress. Unfortunately high levels of stress may block circuits of logical and analytical thinking. The mind automatically focuses on the choice between fighting back or fleeing, which in our civilized life corresponds to aggression or submission.

Assertion

In most cases neither aggression nor submission is the best way to solve a problem. There is a better way to solve conflicts called assertion. This means that instead of shouting back (aggression) or giving up a position you consider correct (submission) you stand firm, just calmly explaining your arguments and your position.

Conflict due to stress

Your counterpart will have a chance to calm down, and you can discuss your different points of view on better terms. Eventually one of you, or both, will have to modify your position. This is now done on the basis of understanding, which has nothing to do with submission.

Summary

A positive attitude
As a manager you are very much dependent on having good relations with your staff members. A positive attitude to yourself and to people around you is the necessary basis for success.

Relationship building
The first four minutes are the most important for building relationships. Make sure you use these minutes as effectively as possible. What you have not achieved in these four minutes is difficult to repair later.

A personal leadership style
Your leadership style should depend on yourself, but remember that your staff members have varying levels of competence and commitment which vary not only from person to person but also from task to task and from day to day. Try to assess competence and commitment and adapt your style of leadership to that.

Motivation
Low motivation in work can be understood and also helped if you have a good personal relationship with your staff members. It is only when your staff members are willing to tell you what their problems are that you can help them.

Pacing
Pacing to adapt yourself to another person's way of thinking and feeling is generally the simplest way to get on good terms with a person.

Stress
Some stress is normal and positive. It keeps us going and helps us achieve our goals. Too much stress can, however, be destructive, both mentally and physically. Identifying and evaluating the reasons for stress is normally the best way to get out of it.

Conflicts

Conflicts can lead to people getting angry. They can also lead to people submitting to the other person, even if they think that they are right. A better way is 'assertion', which means that you control your anger, but also refuse to give up a position you consider correct. Instead you stand calm and firm. Solve the problems using facts rather than with strong feelings.

Good luck!
Good management!
Take good care of your staff.

12. Planning, monitoring and evaluation

Making a good start in a latrine building programme is very important. Equally important is to be able to continue the programme until the project objectives are fulfilled; that is, until everyone in the targeted area has their improved latrine installed and is using it properly. We must therefore consider how to introduce sustainability in a latrine building programme and how to avoid problems in the future.

This chapter briefly describes

- o **the need for planning**
- o **how to monitor progress**
- o **how to evaluate progress.**

Planning, monitoring and evaluation is not a task for experts but is a group task to be carried out by people working on the project

Planning

Up to now, we have addressed different aspects of latrine building, and you should have a fairly good idea of how you would like to implement a programme.

Sooner or later you will be asked to make a plan to implement a programme. Your plan should be based on the kind of technology you want to use and the time you will allow yourself to implement the programme. You also need to assess the availability of the resources necessary to carry out the plan.

In many cases latrine building is carried out with financial assistance from external sources (donations and loans). In such cases there is normally a written agreement, possibly stating what should be done, and sometimes also how it should be done. We are then talking about a project, and the agreement is a project document.[1]

Think of yourself as a planner; expect others to be planners as well

---

[1] Project and programme: A project is a well-defined task, normally defined in a project document, which is a written agreement between government and a funding organization. A programme is

Strategy

Based on what has been said in the previous chapters about target groups, technology, promotion etc., a strategy for programme implementation should be elaborated outlining, step by step, how the programme is to be implemented, by region or district, and giving details of the various activities involved.

Discuss strategy with the people involved...at all levels. And do not forget the women!

Macro planning[2]

The first thing you need to do is to make a general assessment of the whole programme:

You need an overview when you begin...

- which areas and how many families you are planning to include in the programme;
- how many years (or months) the programme is expected to take;
- how production is to be distributed over the project period;
- what the possible financial framework is, how much it may cost and where the money will come from;
- any other given limitations.

A general plan should then be produced, based on your strategy and using the planning data gained from answers to the points listed above.

Detailed planning

Once you have an overview of the programme you can start going into detail with the planning, making up plans year by year and district by district, and deciding where and how you intend to implement the plan.

...next you can start specifying time, people, logistics, money...

Now you need to quantify not only the planning targets but also the resources in terms of material, transport and people. Using this data you should now be able to deduce necessary inputs in terms of human and material resources, transport and finally finance, month by month and year by year.

Micro planning

There is also a need for planning at community level.

...in greater and greater detail

---

a more loosely defined task. Both projects and programmes can be big or small. A project can, for example, be part of a national latrine building programme. A village latrine building programme can, on the other hand, be part of funded projects with conditions defined in a project document.

2 Macro and micro: Macro means large scale while micro means small scale. Macro planning aims to give an overview of the whole programme, while micro planning intends to sort out the details.

Here the programme may be easier to overview, but it is still very common to underestimate the resources required. At community level you will also need to have targets formulated year by year, and month by month, and assess the required inputs in terms of materials, transport, people and money. Micro planning follows other planning, verifying that available resources are sufficient to carry out the programme and are being used efficiently.

Monitoring

The process of planning is about trying to look into the future. Monitoring, on the other hand, is looking at the past to see if what you are doing actually complies with your planning. If so, you have a reasonable chance of achieving what was planned.

In order to plan and to revise your planning you must monitor your progress

To monitor progress, simple and clear progress reports should be made, for example, once a month. To make sure that these reports contain all the necessary information it is very useful to have a special monitoring form to be filled in and submitted to the programme manager as soon as possible after the end of each month.

An example of a monitoring form is given in Appendix 3. Note, however, that your programme may need a different form with perhaps more or less information, specific to your monitoring needs.

Revisions

If you compare your monitoring results with your plans, you will usually see that your plans can be improved. For your planning to become more accurate, you have to revise your plans at intervals, possibly once a year, or when you, as a manager, feel that the old plan no longer serves its purpose.

Evaluation

Everybody involved in a programme should make it a regular habit to ask themselves if they are doing the right thing in the right way. At programme level a more systematic evaluation of progress is required from time to time.

Everybody should be an evaluator...

When correctly done, such an evaluation should be made with eyes open for any possible improvements. It is

...looking for improvements

therefore common to invite external evaluators to participate in the evaluation. These evaluators should be seen as helpers or assistants in the programme, and not as police officers. If they do their job in the right way, you should find that they will help you to solve the problems you have, rather than giving you new ones.

In sanitation programmes it is very common to find that progress does not follow the original plan. This may be comforting to know, even before the evaluators arrive.

In some cases the evaluators may find that records are not being kept in order, that materials have disappeared without being reported, etc. It is the duty of the evaluators to investigate all such irregularities. To avoid these problems you should be careful to report not only your progress, but also your problems, and always to keep your records up to date.

Report problems and irregularities on a regular basis

What should be evaluated

There are three important aspects which come up each time a project is evaluated. These are:

- affordability
- acceptability
- sustainability.

The evaluators should also look at how the programme is being implemented and if it is receiving, and using, its resources in the proper way.

Normally an evaluation is made against the plan and its formulated objectives. During the evaluation, there should be the possibility of discussing the objectives and targets of the programme. If not, the evaluation process is in danger of becoming mechanical, and important aspects may be overlooked.

Discuss objectives and targets critically

Affordability

We earlier addressed the problems of cost and economy, and the importance of proposing latrines that people can afford. We have stressed the importance of the low cost of SanPlat latrines when compared to other improved latrines. However, there may still be financial problems for the lowest income group in meeting the cost.

A difficult balance

There are various ways of reducing costs, but great care must be taken, as cutting costs may create new problems. The following proposals can be used separately or be combined:

1. Change the design
Ventpipes, water seals, concrete slabs, sheet metal roofs etc. are expensive and not always necessary.

Look for affordable solutions

2. Reduce the price through increased subsidies
Reducing the price through increased subsidies may be an effective method if the subsidy can be maintained after the introduction of the programme, or if it is anticipated that the subsidy will only be required at the beginning of the programme.

Subsidies may be dangerous if sustainable funding is not secured

3. Reduce costs through training and competitive bidding
Private production is known to be cheaper than public production, but only if there is competition. Private producers can be trained and supported in practical ways to produce SanPlats and improved latrines. Orders may then be placed after competitive tendering with quality control. No payment should be made for SanPlats not complying with the specifications.

Note that quality specifications and a reference sample should exist, and be referred to when the contract is signed.

4. Introduce a market-defined floating price
Allow sale prices to be low at the beginning of the programme and allow any loss be compensated for later, when the demand has risen to a level where prices can cover initial losses. Any destitute families, and other people who have not improved their latrines, should then be taken care of in the final phase of the programme, so that 100 per cent coverage is achieved.

Let the good times pay for the more difficult periods

Warning: It is a common mistake to combine experimental and demonstration latrines with social charity by providing latrines at no cost to destitute households. The result of this will be negative, as many people will associate the new low-cost latrines with poverty and low status. Demonstration latrines should be built for, and by, the EER-group (the Educated, the Employed and the relatively Rich), and there is no reason to give any special subsidies for this group. On the other hand, they should have what the rest of the community can expect.

DANGER!!

In order not to jeopardize the promotional programme, the social aspects of latrine building should wait until later, when the social acceptability of the programme is secured.

Acceptability

We have also stressed how important it is that the latrines are liked, and approved, by the population and by the people promoting them. If people do not like the latrines you may have chosen the wrong technology. Questions such as those listed in Chapter 9 should form the basis of some initial market research to help avoid this. Acceptability may also depend on the way in which the new latrines have been promoted. Reveiw what has been written about promotion in Chapter 10 and about how to work with people in Chapter 11.

Sustainability

Sustainability means that the implementation strategy you have chosen should be possible to implement, not only at the beginning of the programme, but also in the future, when the initial enthusiasm has faded and when other programmes and priorities are also calling for attention and financing.

Who is taking care of the programme at the end of the project?

Even if you have this in mind when planning the programme, things may change over time. The objective of monitoring and evaluation is to try to identify problems at an early stage – before they become serious – and to suggest possible solutions.

Key questions

In the evaluation of a low-cost sanitation programme there are a number of questions which are critical. Some of them are listed below. There may be other questions which are equally important and sometimes even more relevant, as each project has its own profile and its own problems.

Discuss problems and opportunities with the population and your staff at all levels

General

What is the principal obstacle facing the programme?
Often there is another problem in the community which is considered to be more important, and which demands the people's and the leaders' attention. Water may be such a problem, agriculture another. This is one of the reasons why sanitation may be best integrated with other projects, assisting the community to solve a felt need.

The lack of general understanding of the need for latrines is a common problem which can usually be addressed through health and hygiene education. To become effective, hygiene education takes time and needs good latrines already installed. That is why we often recommend the promotion of latrine construction to precede hygiene education if the two cannot be combined.

Hygiene education is important but takes time

Promotion which uses the full range of arguments including privacy, status and convenience as well as hygiene and health will be the most effective. The concept of being 'modern' has the reputation of being the single most effective argument.

The new SanPlat (cast in a plastic all-in-one mould) is more modern than 'traditional' SanPlats: see Appendix 2

The SanPlat system has been developed to be low cost and affordable by all. Has this been the case here? What strategy has been used to reach the poorest families and the uninterested ones? Should SanPlats be for all, or only for those who can afford them?

The SanPlat system has been designed to make subsidies effective, as a small subsidy can have a considerable impact on the price. Has this been the case here? Has it been important for the implementation rate of the programme or has the reduced price lowered the status of improved sanitation?

Are we building the right type of latrines? If not, why? What can be done to improve the technology?

Warning: Improving the technology often implies that the product becomes more expensive, making it less affordable. A subsidy is often limited to a certain amount per family. Increasing the cost of the technology may mean that a bigger portion of the cost needs to be carried by the household. If we instead reduce the cost, a bigger portion will be carried by the subsidy with a smaller portion to be paid by the household, thus encouraging increased community participation. However, if the subsidy is increased the money available may not be enough for all. There is a need to find a balance.

Have you used 'market research' to find out about alternative latrine designs and possible costs?

Are the latrines we produce liked by the people? Can they be modified to become better still?

If transport is a problem, the latrine slabs can be made in segments. If there is a felt need for concrete slabs, SanPlats may need to be integrated into the slabs, and the slabs cast on site.

Can they be made even more simply without affecting acceptability?

The introduction of handcarts for transporting slabs

proved to be very successful in the Mozambican latrine building programme.

Are the new latrines well-used by all? If not, who is not using them and why?
Children are often in the difficult user group. Chapter 10 gives some useful hints on how to encourage children to use latrines. Another problem may be that the latrine is too far away for the smallest children. Perhaps they should have a very simple toilet for themselves closer to the house.

It is the use that is important, not the design

How can latrines be improved ?
Squatting may be a problem for people with stiff legs, elderly people or pregnant women. The solution may be a raised SanPlat for sitting or a handle to hold on to while squatting.

The SanPlat system has been designed to fit into existing latrines. Has this been the case in your project? Have there been any special problems and how have they been solved?

Many latrine building programmes have failed because they have tried a technology which is too complicated or too expensive. The SanPlat programme is designed and tested to be as simple as possible while remaining acceptable to people. Has this objective been fulfilled in your case?

Planning

Are there plans for the implementation of the programme?

Are the targets well set? Is it possible to carry out the plan with the existing resources, with the existing strategy?

Is the plan being followed? Is the plan subdivided, so that the individual officers can see if they are doing well enough or not?
If not, why? Is the plan lying in a drawer or is it on a wall?

Is the plan made up as a graph or is it only 'dead numbers'? Are the achievements plotted on this graph as well?

What is likely to happen if the plan is not followed?
Depending on the commitment of the politicians, plans can have differing levels of status. Some plans can have almost the same status as a law. But, when correctly

used, a plan should be thought of as a tool which helps you do your work – always ready at hand when needed.

Have the targets been set considering the total need of country, district, area, etc? Will there be improved latrines for everybody within a reasonable time? If not, why?
In the initial stages the plans may be vague. Eventually, however, there is a need to plan for the total coverage of the whole area.

Has the community participated in the planning of the programme?
For a community to achieve total coverage there must be some level of commitment from the community leadership to follow the plan.

Does the implementation of the plan imply that the community participates in one way or another? Could the planning have been improved? In what way?
Do the plans bring any other benefits to the community (such as water or health education)? What is likely to happen to these benefits if the community does not fulfil its part of the agreement?

Were the right people involved? If not, why?
Local leaders may be formally appointed, but others may be more influential. What could have been done to get more people involved in the planning?

What could have been done differently during the planning stage?
The SanPlat system has been developed in such a way that it is possible to implement together with other programmes. Has this been the case in this programme? If not, why?

Would there have been any positive advantages in integrated implementation?
Or was the programme originally integrated, but suffered from a lack of priority? What is the project management's idea about that?

Have you agreed on some form of community participation?
For example, have the benefits of the provision of water been made conditional on the construction of latrines before starting well construction?

Is the project serving all equally, or is priority given to villages which comply with signed agreements?
It may be difficult and unpleasant to abandon a village which has not complied with what has been agreed, but priority should be given to villages which have complied with the agreements.

Is there any possibility for a certain village to receive priority service if it increases its community contribution to the programme?
It may be a good idea to compare the kinds of priorities recommended for individual households when starting up the programme. Also, villages could be given priority if they commit themselves to a higher level of community participation. Eventually villages may end up in competition, each trying to do better than the others. In fact they will all be struggling to achieve better living conditions for themselves.

Implementation

How much coverage does the programme have today?
Who is making the latrine counts and when? Is information based on official statistics, or have the programme staff been around checking for themselves? How has the estimate been made? Is it a total count or an assessment based on samples?

Your monitoring reports should be informative

How many families do not have improved latrines?
Surveys and latrine counts always have a mobilizing effect. They are evidence that somebody cares. Many families will have learned about the new latrines, others may need more information. Having people in the field should always be used as a way to improve the information level.

Simple surveys make good use of time and money

Has the programme been integrated in theory, but not in practice? Are there other components which are considered to be more important, or which have a different rhythm of implementation?
Many water and sanitation programmes are water programmes with a sanitation component tacked on because the donor wanted it. We must remember that an improved water supply alone will not give health, nor will improved latrines alone. It is only when there is a combination of clean water, good sanitation and sound hygiene practices that improved health will become a reality for the community.

Integration has both positive and negative aspects

Do the people involved have any suggestions?
We very often talk about the importance of listening to people's suggestions, but what actually happens to the answers? Is there a dialogue between the programme management and the population – or are we, in fact, telling people what to do? Do we collect people's opinions on forms which become dead heaps of paper? There may be other ways of reaching people and sharing the programme implementation responsibility.

Are the local leaders happy with the implementation of the programme?
When signing an agreement with a community, make sure that you are not signing for more than you know you can deliver. Be sure to explain any risk factors which may be involved in the programme at an early stage when discussing the agreement with community leaders. Equally, you should also try to ensure that the community leaders do not sign for more than the community will be happy and able to comply with.

Are your staff members happy?
Your staff members are your most important asset in the implementation of your programme, and internal staff problems could easily jeopardize the whole programme. Do not allow such problems to develop – deal with them promptly and quickly! More information about how to solve problems of this kind can be found in Chapter 11.

Your workforce may have the information you need; ask them first

Many of your staff members may have very close links with the community. They may live in the area and meet the people each day. What do they think?
A practical way of pulling out problems and finding solutions is by having regular meetings with your staff, where they can list and discuss any new problems and develop solutions. A blackboard or a flip chart can be extremely helpful in such meetings, as it helps to organize information and make the participants aware that their problems and solutions are not forgotten.

Do you foresee any problems for long-term implementation of the programme, to reach its target of 100 per cent coverage?
In many programmes the availability of the SanPlat has encouraged the construction of more and better latrines. Has this been the case here? If not, why?

Production

The SanPlat has been developed for possible local production.

Have SanPlats been produced locally? If not, why? Would that solve any problems? Would it create any problems?

Quality

Does the quality of the SanPlats comply with stated SanPlat standards?

Experience from many projects shows that the quality of the SanPlat has an important impact on people's interest. More than that, the general appearance of the SanPlat has a positive effect on the cleanliness of the toilet. These are aspects which may seem less important at the beginning of the programme when the SanPlat concept is new and when 'everything' sells. When the first enthusiasm is gone people become more critical.

SanPlats that look pleasing sell better than ugly ones and improve your credibility

Has production kept pace with demand? If not, why?

In many countries the availability of cement is a bottleneck. Transport may be another. In successful programmes it may be the production and control capacity which is the limiting factor.

Do not promote more than you can supply

Have you tried to modify the production process in order to make it better able to cope with the demand?

The introduction of full-size plastic moulds may be a solution. Production is easier and faster, and quality improves. Less training and control is required and production becomes more fun. A disadvantage may be the increased cost for the moulds, but this may be marginal, or even negligible, if a great number of SanPlats are to be cast with the moulds.

Economy and financial management

Money is always a hot issue. Money may be abundant at the beginning of the project but almost non-existent by the end of the financial period. Or it may be that some money was not spent during the first year, so the budget was cut for the second year, and expansion of the programme is no longer possible.

Not spending allocated money may be seen as a lack of progress by funders and may affect your budget allocation for the following year

Is there a plan for expenditure?
It is necessary to regulate spending according to the plans?

If savings result in plans not being implemented, they

may have a negative impact on the programme. Project resources are allocated to be used for the project.

How does the project receive the payment from its customers?
If people want to buy but there is nobody to receive the money, something may be seriously wrong. If the demand is strong and everything is sold it may be wise to concentrate sales to certain hours of the day, or to a special day each week, as this will limit the problem of cash control.

Payment for goods is a common bottleneck

How are records being kept?
Theft is a very common problem in projects, and money is easily stolen. Once this has started there is a serious risk of more and more money being stolen. Simple and straightforward records of stock, production and sale are the absolute minimum requirements. A diary, where all production, sales and stock records are kept and recorded daily, is a good way to maintain control. The number and date written on all slabs should be registered in a daily production record. Receipt books should have pre-printed numbers and carbon copies.

Do people have confidence that all the money is being taken care of?
Just the suspicion that somebody may be stealing is demoralizing. All records should therefore be checked on a routine basis each week, and care should be taken to explain the checking system to all staff members at your regular meetings so that they are aware of what is going on – and why. Ask openly if this system of recording and checking is a good way to ensure the security of cash and income from sales etc.

Money is a sensitive issue

Simple solutions are often the best ones

If not, what improvements can be made?
If there are problems, look for methods of improvement, but be careful not to make the systems complicated. The more complicated they are the more difficult they are to control and the greater the risk of confusion and theft.

Promotion in schools

Are the local schools involved in the programme?

What are they doing?

Could they do more?

What children learn in schools is passed on to younger sisters and brothers and other children, even parents

If so, what?

Which teachers are doing what?

Which teachers are not participating – and why?

Which pupils are participating – what are they doing?

Which pupils are not participating – why not?

Promotion in religious institutions

What are the churches and other religious institutions doing?

Which churches and institutions are involved?

Could they do more?

If so, what?

Religious leaders often have more contact and influence than the elected political leaders

People

Who is involved?

Are the right people involved?

What are the criteria for determining who are the 'right' people?

What is the role of women and children?

Who is doing what?

Who else should or could be involved?

Are people in local households involved in the programme?

How are they involved?

If not, why not?

Can the participation of ordinary people in the programme be improved?

If so, how?

There are special methods for involving people; they have abbreviations such as SARAR, PROWESS, PHAST and RAP

Women

What role have women played in the programme?

In planning?

Would it be appropriate to involve women in the planning of the programme? What are the consequences of not doing so? What do the men say about including

Women have power to promote ideas. They know how to talk and convince. They will become still more effective if involved in planning, monitoring and evaluation

women in the planning committee? What do the husbands say? If involving women is a problem, what is the solution?

In construction?
Do women build latrines? Can they do it? What would be the advantages or disadvantages in involving the women?

In promotion?
Women are known to be the first ones to understand the need for improved latrines. Has this been used in the programme?

In monitoring?
Women monitoring how many families have and how many have not improved their latrines or built new ones gives many opportunities for them to take part in promotion. Women talking to women is more effective then men talking to other men's wives. Remember that monitoring should be made in a promotional and evaluative way.

In evaluation?
If women participate in the evaluation, women's points of view are more easily included in the programme. This may be critical in your efforts to reach out to female-headed households.

Monitoring

Is progress being monitored on a monthly basis?

How?

If not, why not?

Is the monitoring form easy to understand?

Who is receiving copies of the monitoring reports?

Could monitoring be carried out in any other way?

Evaluation

Has the project/programme been evaluated before?

If so, what were the principal problems which were addressed in the previous evaluation?

Who has been involved in the evaluation?

How did they participate?

Can women be involved in building?

We can close the circle by monitoring and evaluating the monitoring and evaluation process

Evaluations must also be evaluated...preferably by staff members and the community

Have there been any improvements since then?
If not, why not?
Should the evaluation be made in a different way?
If so, how?

Appendix 1. Latrine designs

This appendix is a collection of different latrine types. All the latrines include SanPlats, but they may also be classified as varieties of VIP latrines, pour-flush latrines, and so on, and are appropriate to different sites and circumstances as indicated.

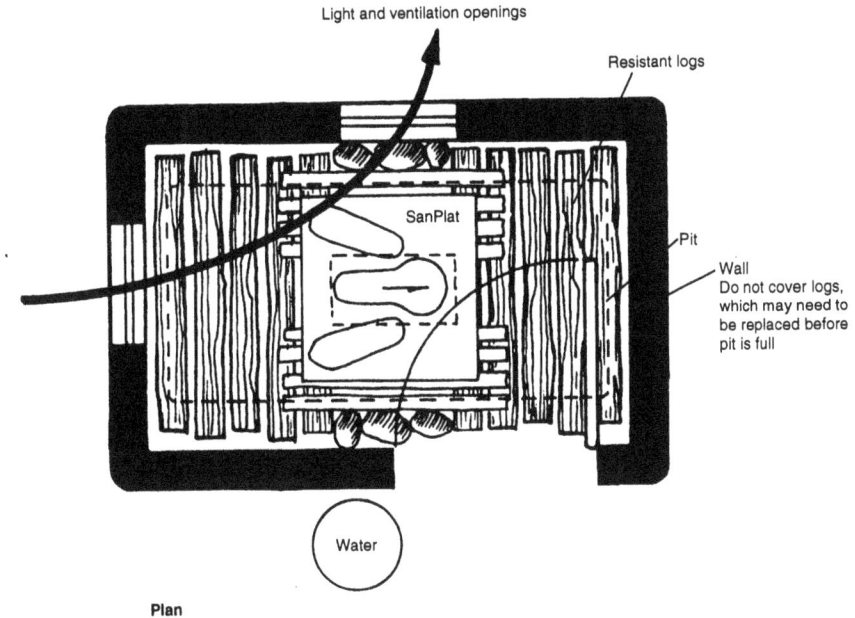

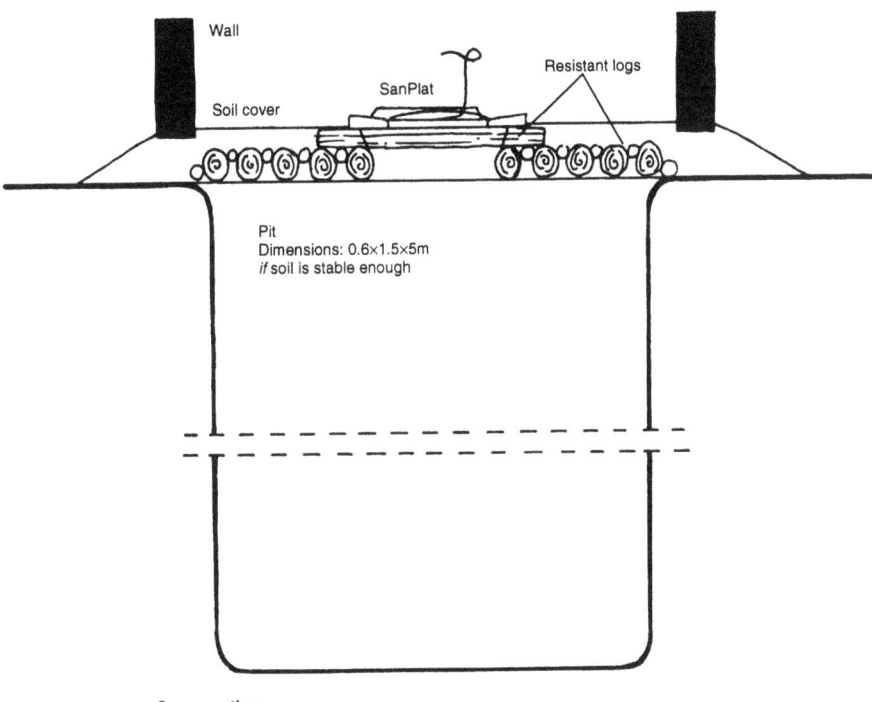

SanPlat latrine with resistant logs

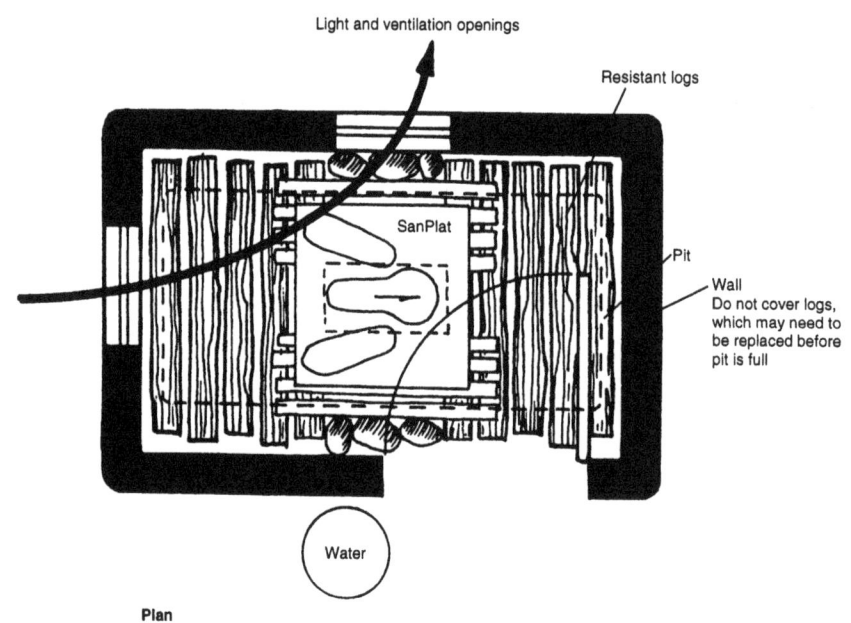

Elevated SanPlat latrine with resistant logs

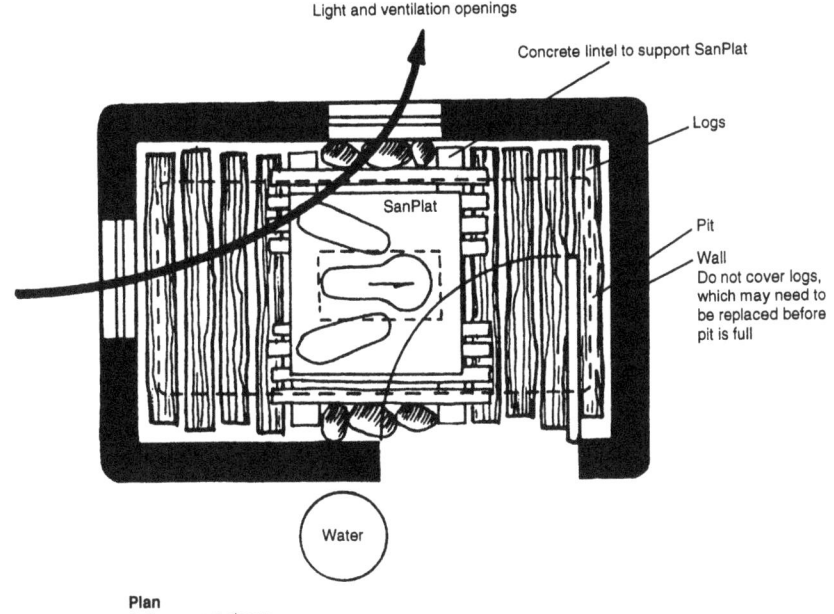

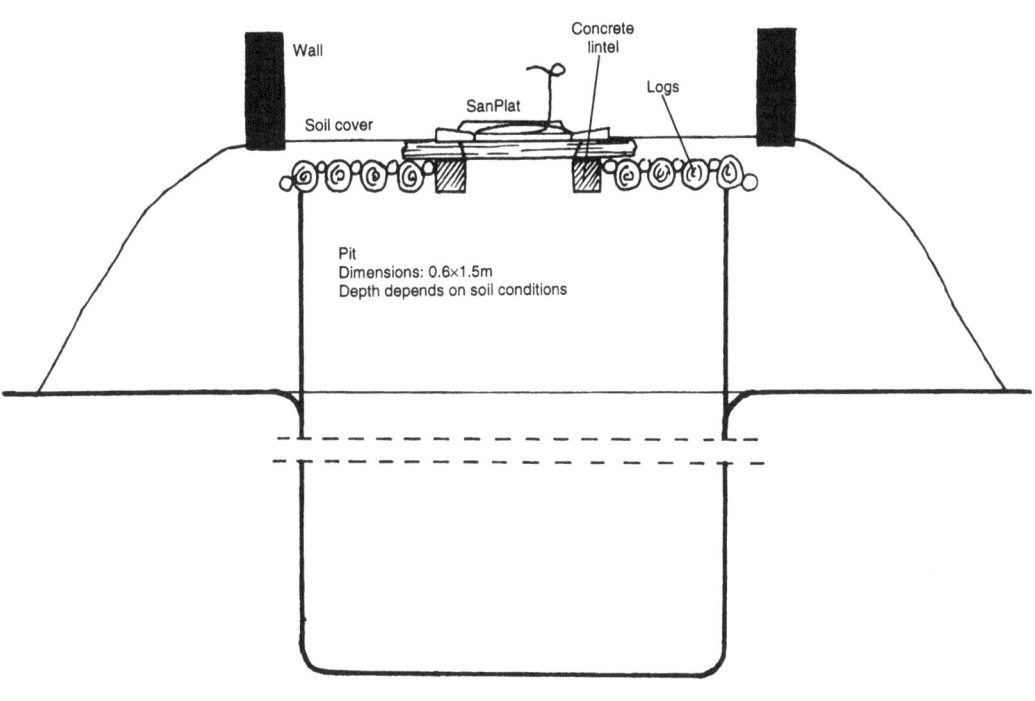

Elevated SanPlat latrine with logs and lintels

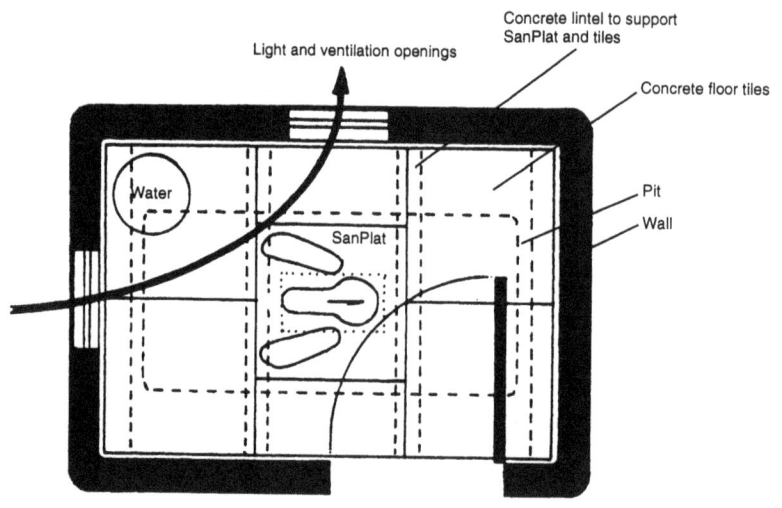

Plan

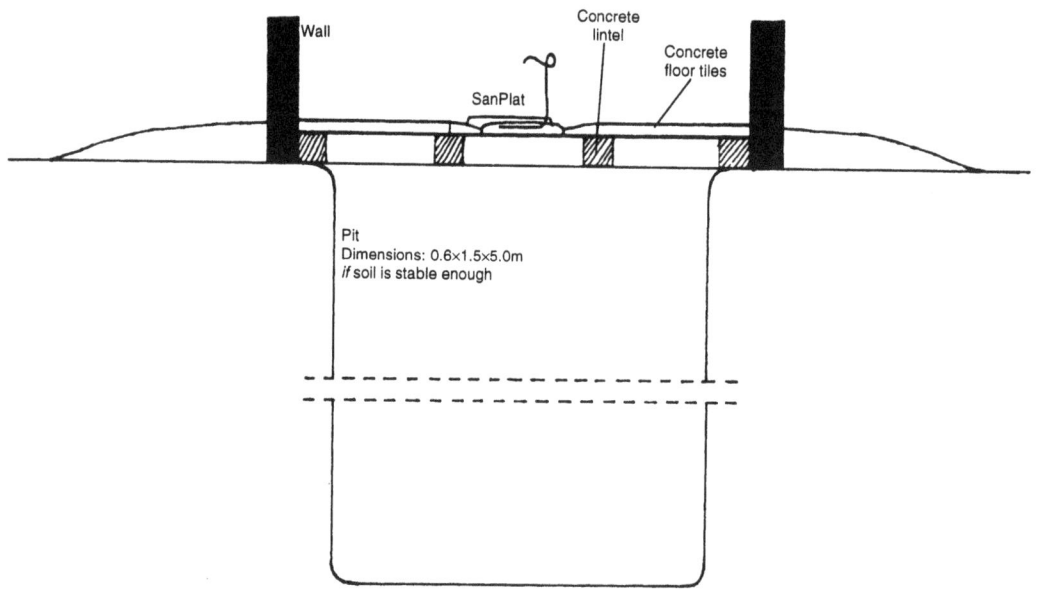

Cross-section

SanPlat latrine with lintels and tiles

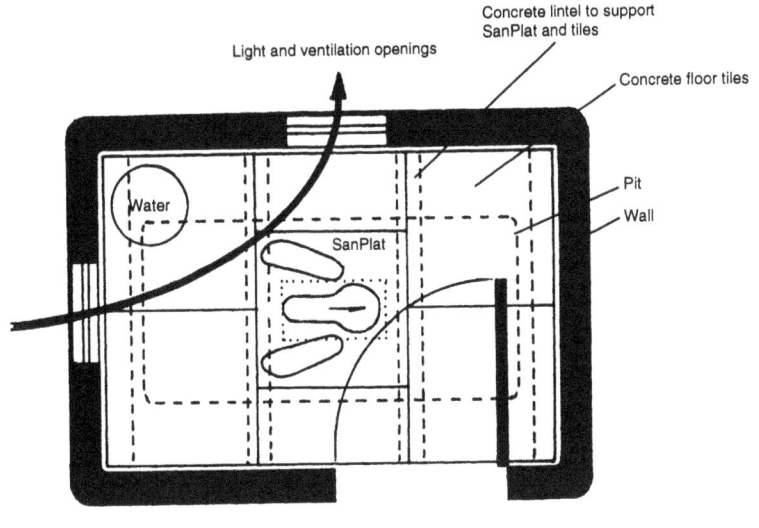

Plan

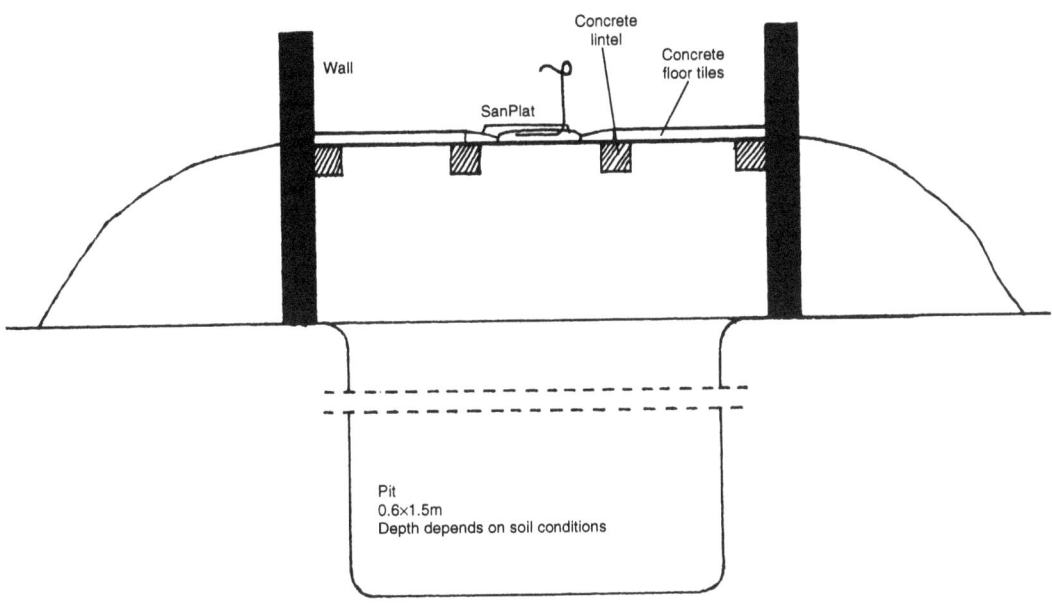

Cross-section

Elevated SanPlat latrine with lintels and tiles

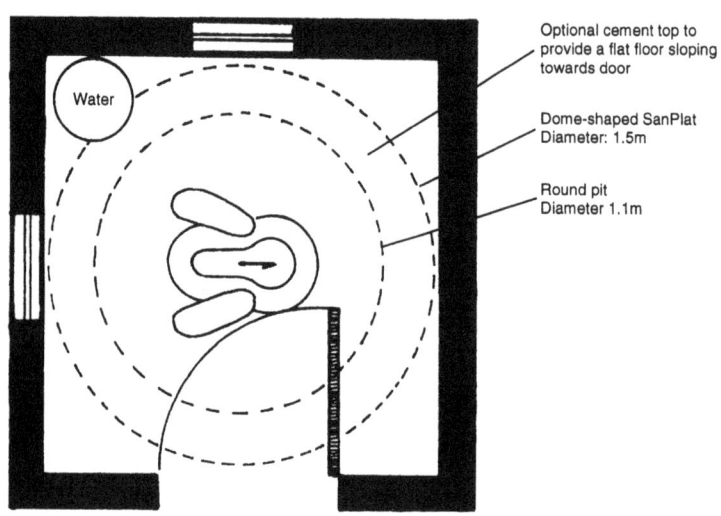

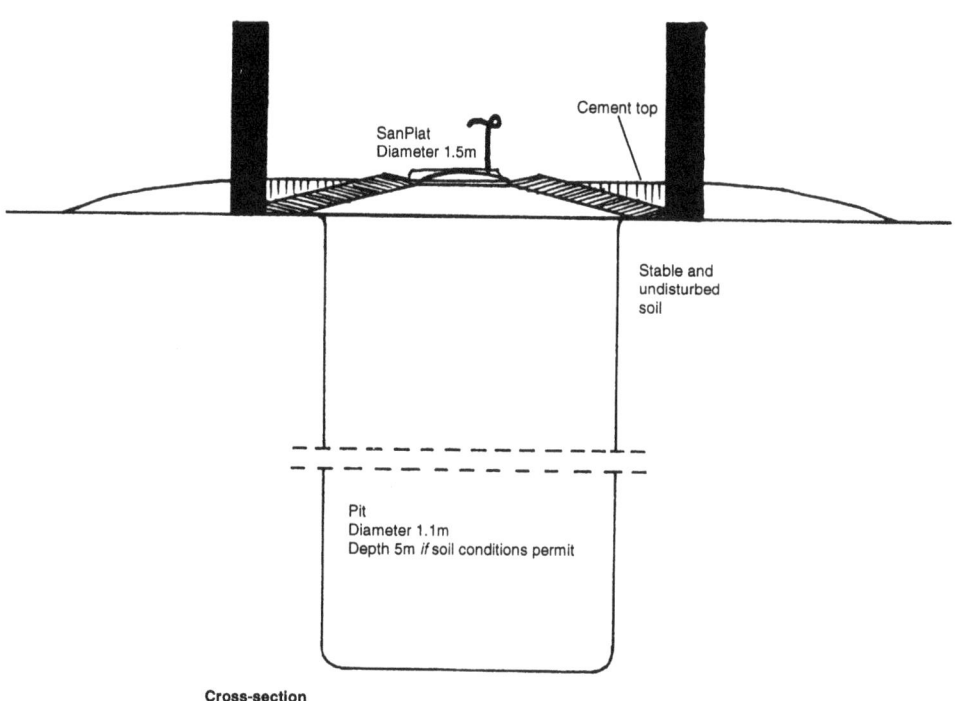

Latrine with dome-shaped SanPlat for stable soils

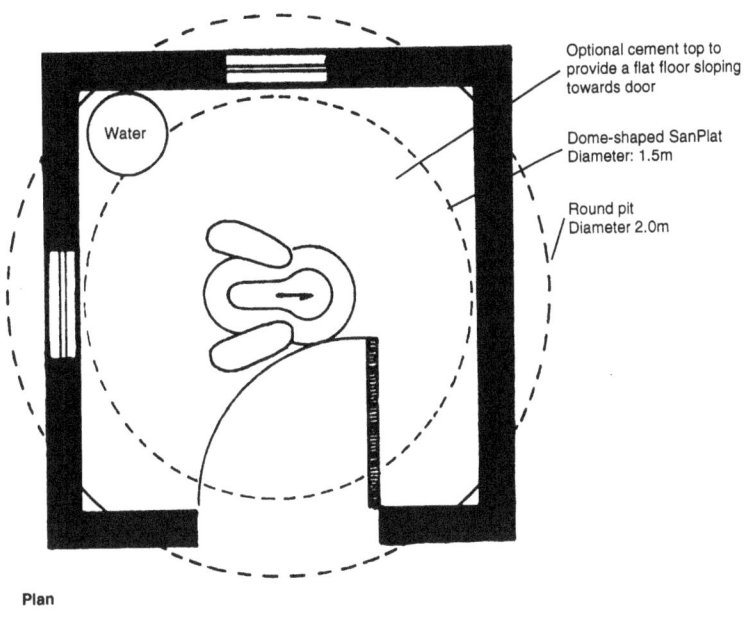
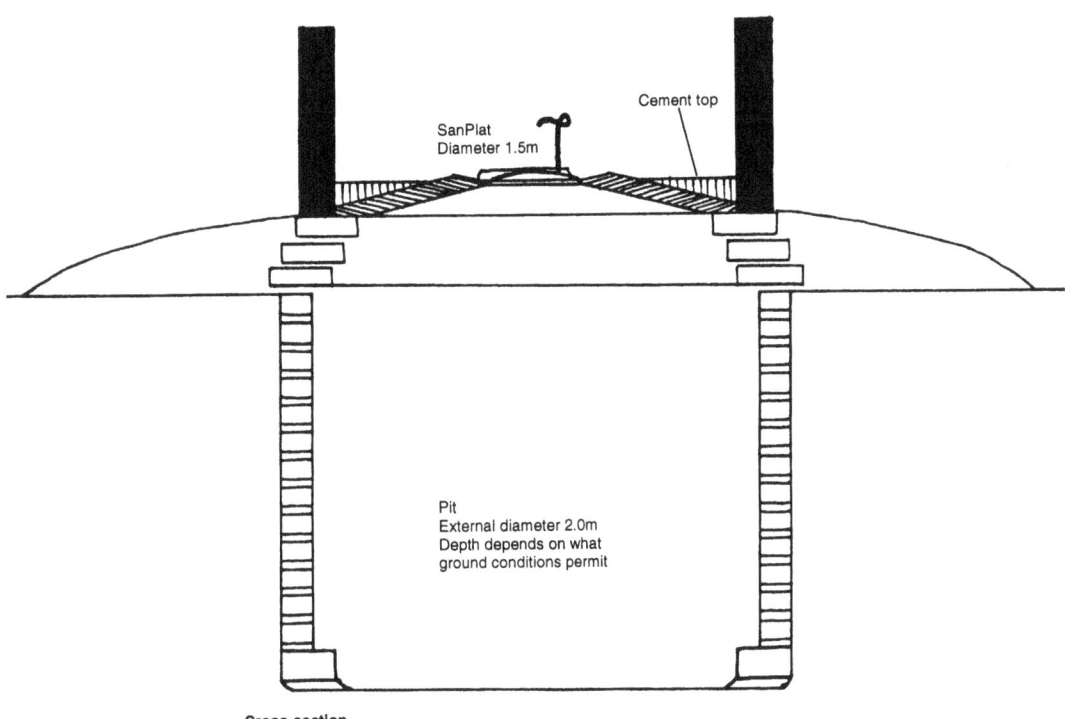

Latrine with dome-shaped SanPlat and lined pit for unstable soils

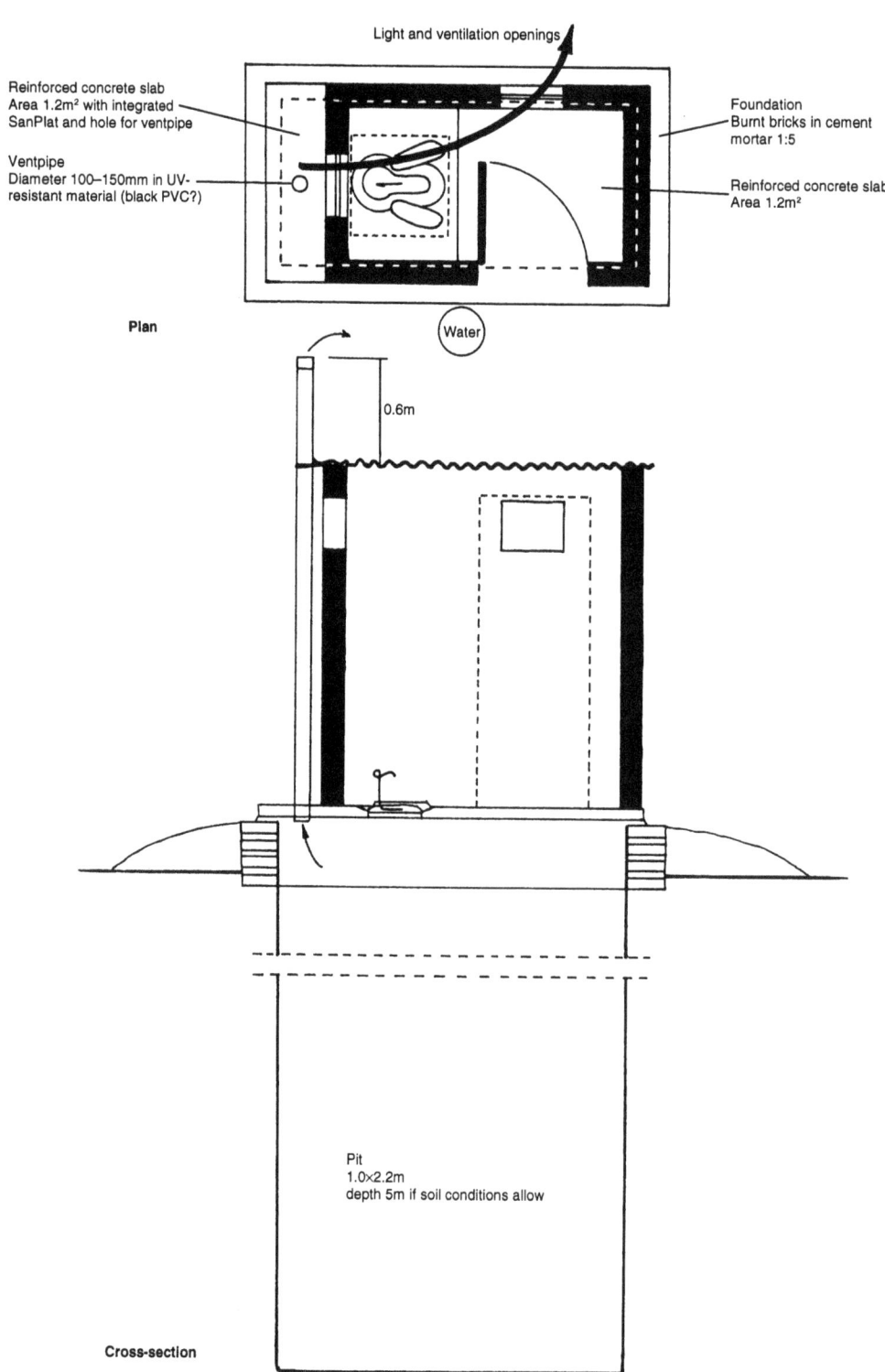

VIP latrine with standard SanPlat for stable soil

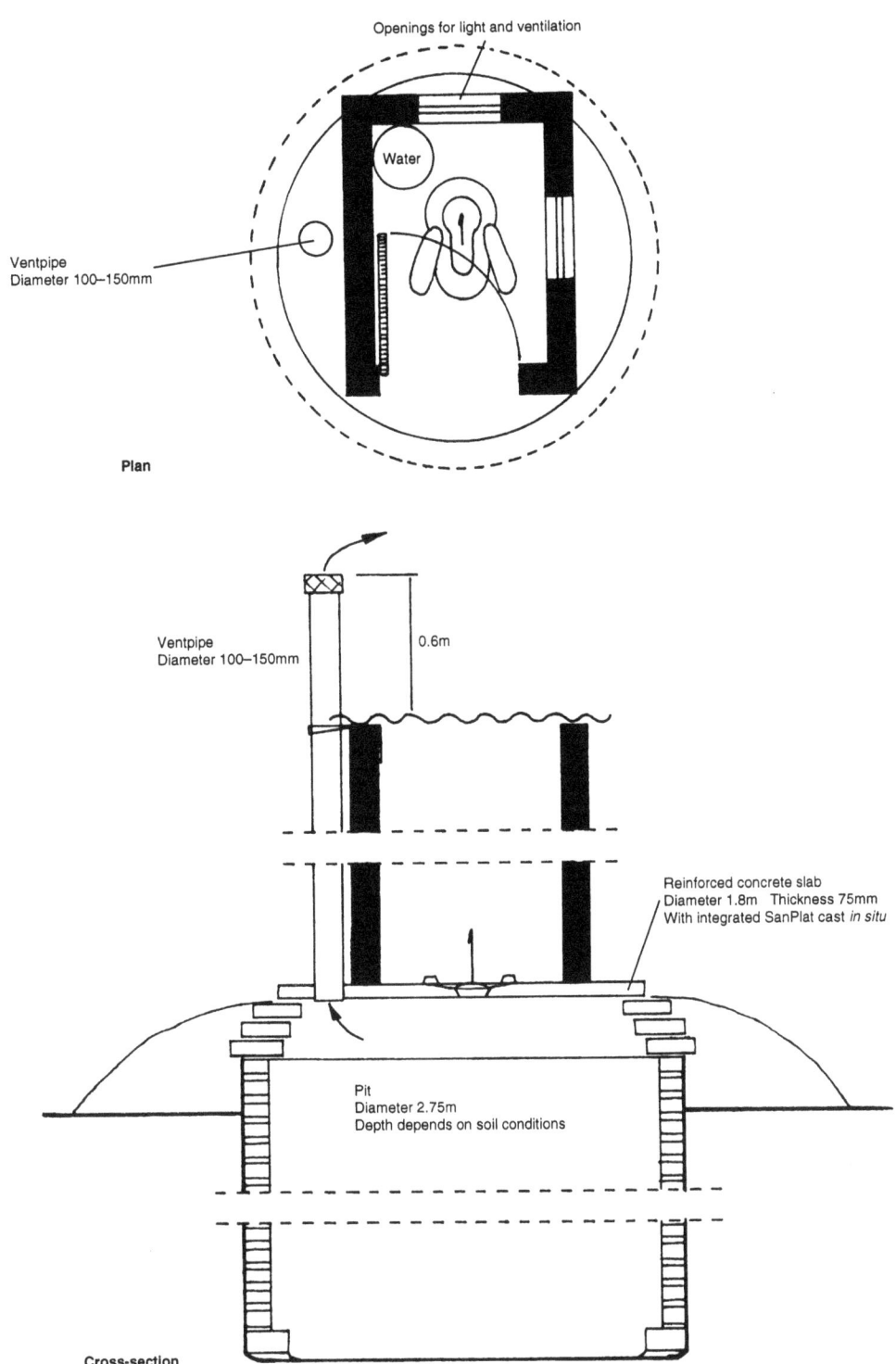

VIP latrine with standard SanPlat with round lined pit for unstable soils

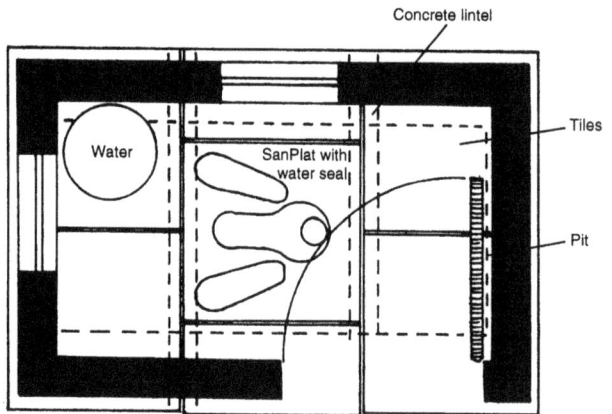
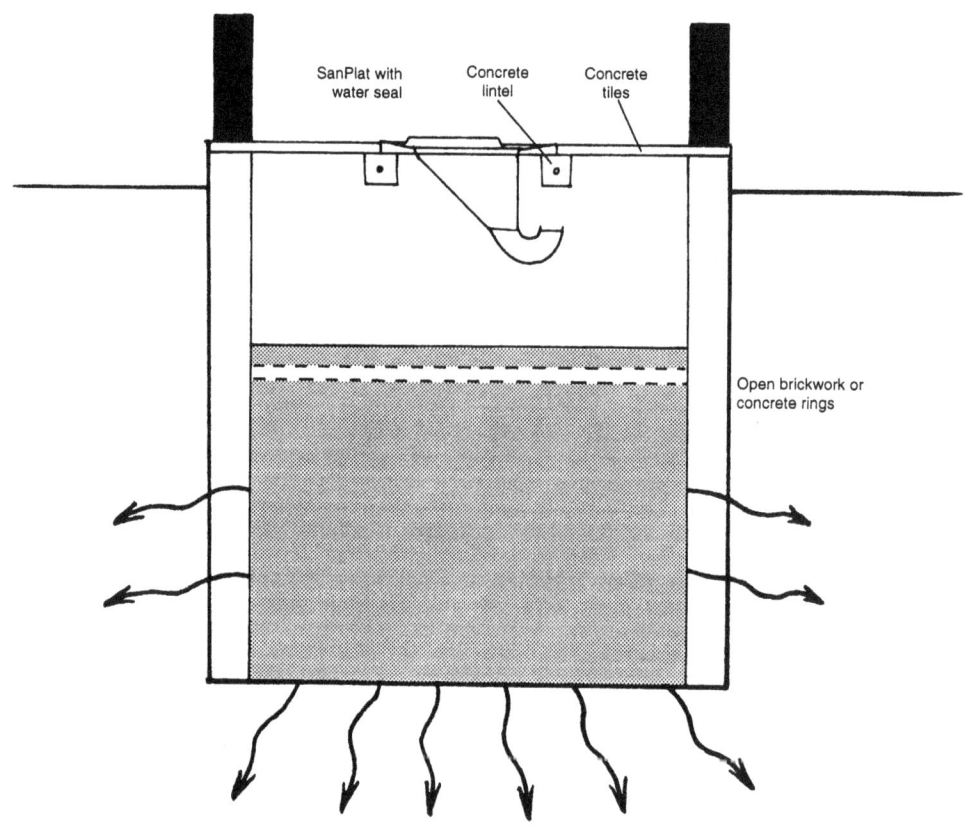

Pour-flush latrine with direct pit

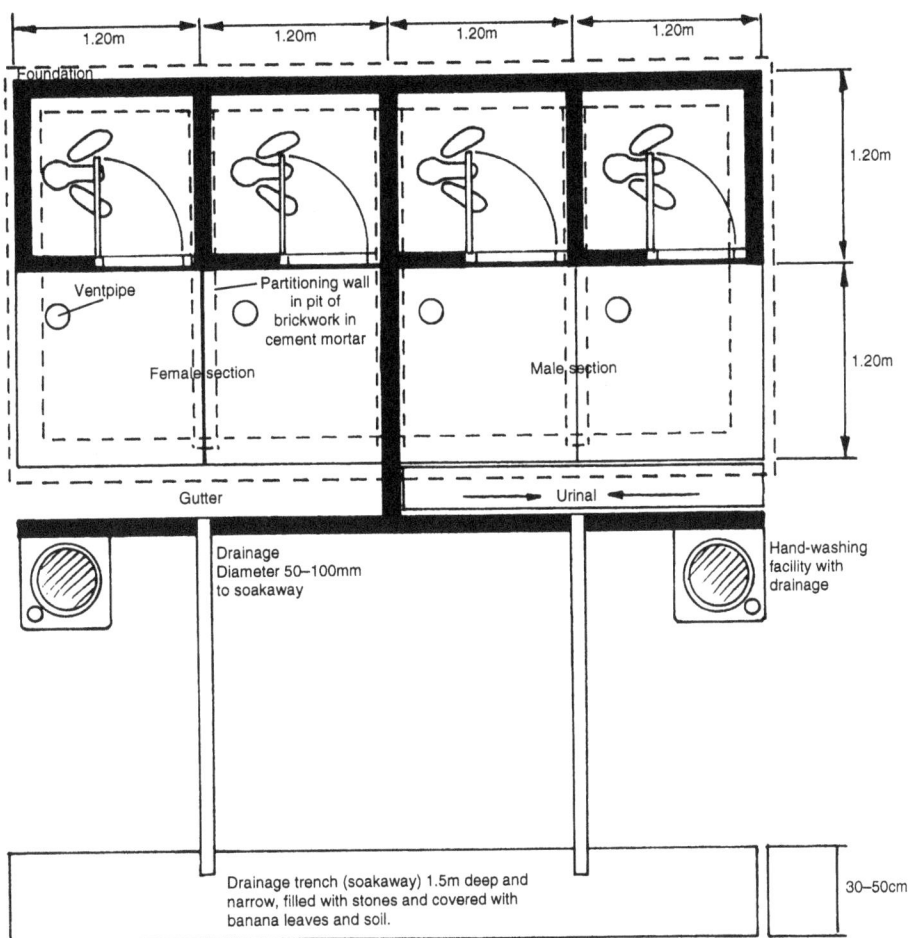

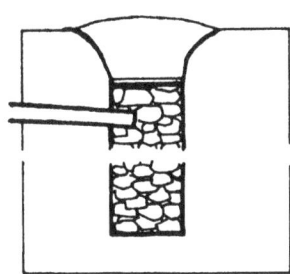

Drainage trench cross-section

Institutional latrine
Drawing shows principle; number of compartments depends on number of users

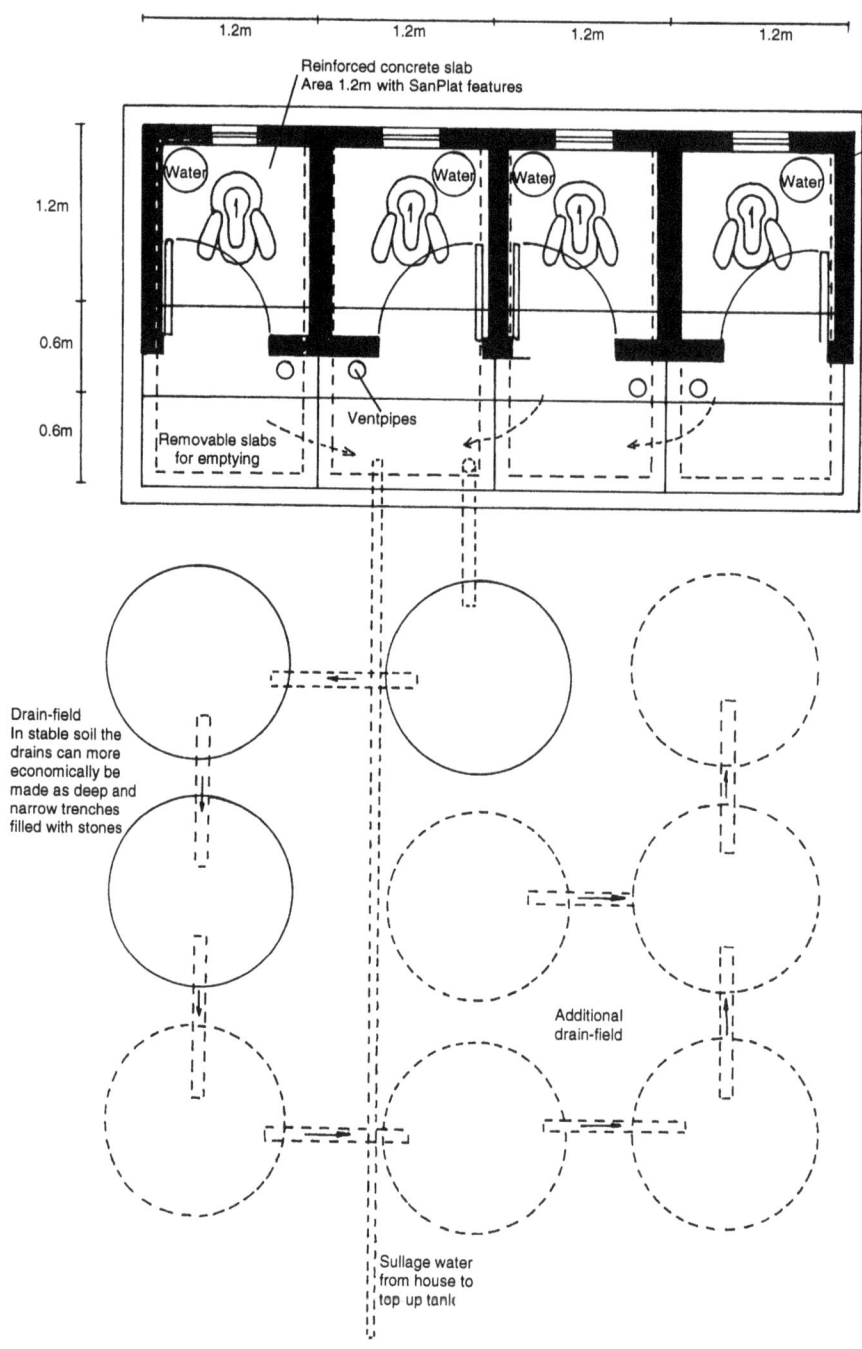

Septic-tank-based VIP latrine
Two square slabs 1.2×1.2m² per compartment (Plan)

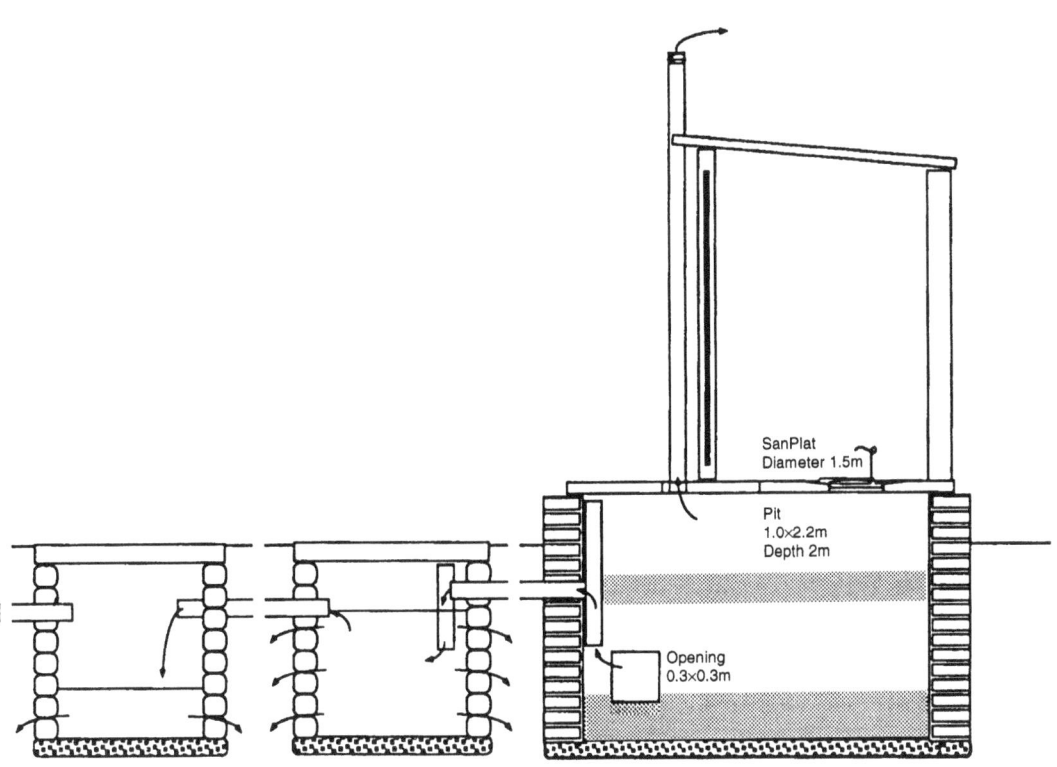

Septic-tank-based VIP latrine
Two square slabs 1.2×1.2m per compartment (Schematic cross-section)

Appendix 2. Casting the small SanPlat using the all-in-one plastic mould

This method for casting the small SanPlat was developed after the book was written. We have therefore included the instructions as an appendix.

1. Start by greasing the shiny side of the all-in-one SanPlat mould with clean motor oil or grease on a cloth. The oil will protect the mould from the cement.

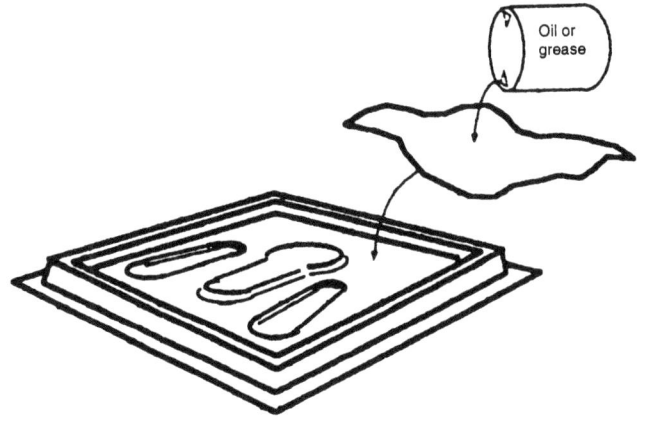

*Grease **all** surfaces including middle and outside*

2. Wash the sand and stone if you have any doubts about its purity and let it dry before mixing with the cement. Dust and impurities make the concrete weak.

   Mix cement with clean and dry sand and stone in the proportions 1+2+2 (or 1+2+3), add very little water and mix well.

   Divide into two parts and add extra water and cement to the softer of the heaps to make it liquid. Stir well in a bucket or laundry basin.

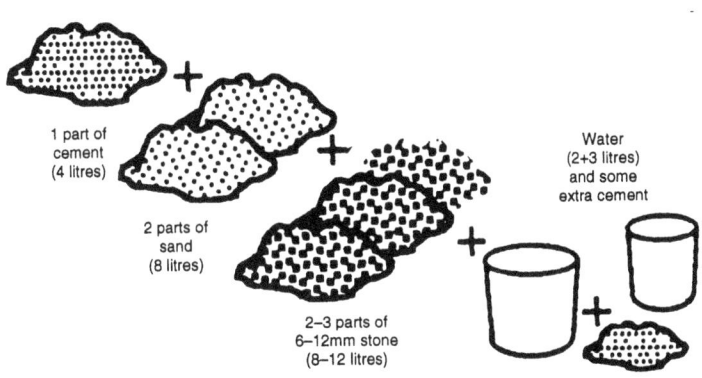

*Use **clean** materials*

3. Pour the liquid concrete into the mould and stir gently with a trowel or a twig. Hammer gently on the mould to release all air bubbles from the plastic surface. Be careful not to scratch the mould.

   Put in reinforcement bars as required before adding the very stiff concrete to cover the reinforcement well. Extra reinforcement may be required for transport reasons.

   Allow the dry concrete to soak up the water, and compact the concrete with a piece of wood until cement water comes up on the surface. (You may choose to start another slab while the water soaks up.)

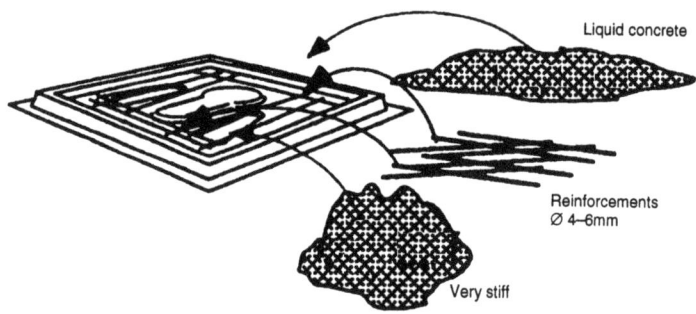

Fill with the soft concrete first

4. Write the date and the number of the SanPlat in the fresh concrete. Start with 1 and continue with 2 on the next one, and so on. You may need to sprinkle some cement at the place of writing. The date and the number will help you monitor production and progress.

   If it was too wet, use less water for the next one.

   Put the mould on a flat surface and allow to harden for one or two days. Using two planks under the foot-rests, you can put the finished SanPlats in a pile and save space.

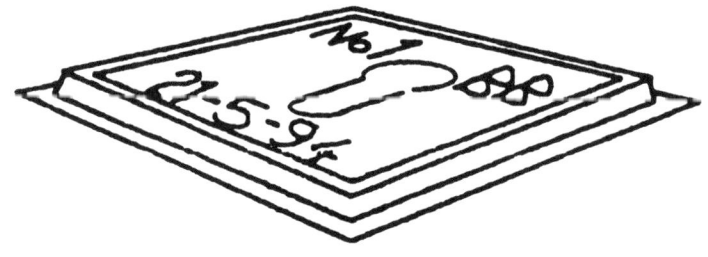

Write the inscriptions on the back

5. When the concrete has set, turn the mould very gently over soft sand or grass to get the SanPlat out of the mould. Do not let it fall! It is still very weak.

   Clean the mould with a soft cloth and some oil and it is ready to be used again. Do not use sharp objects or stones to remove concrete remains.

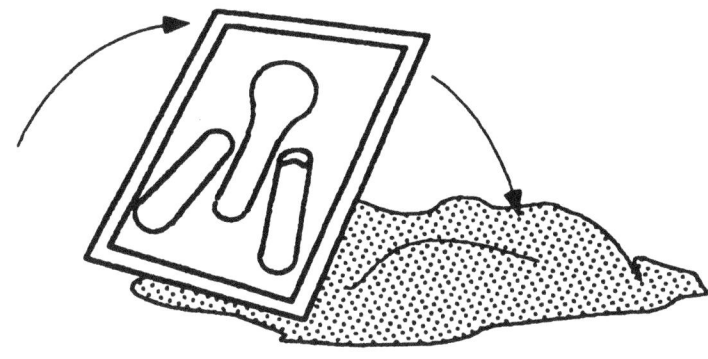

Turn... but do not let it fall!

6. Put the SanPlat (and the moulds) in the shade so they will not catch too much sun, cover with plastic and sand and keep the SanPlat wet for one week.

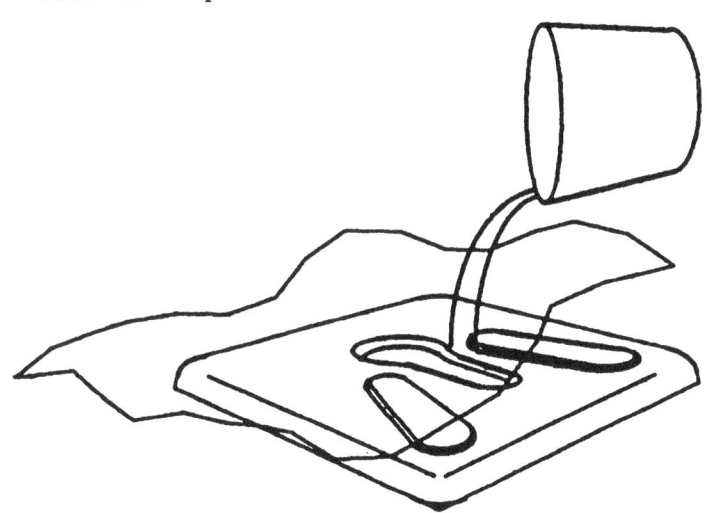

*Keep it wet for **one week***

7. A concrete lid can be made using a special mould, but you can also use the back of the all-in-one mould as a mould for the lid.

   A piece of wet paper (from the cement bag) or some oil will prevent the SanPlat from sticking to the fresh cement. The handle can be made from a piece of reinforcement bar fixed in the concrete. Put in extra reinforcement to make it extra strong.

   Use a piece of a polythylene bag to smooth the surface and keep it wet for one week to cure properly.

   Ready!

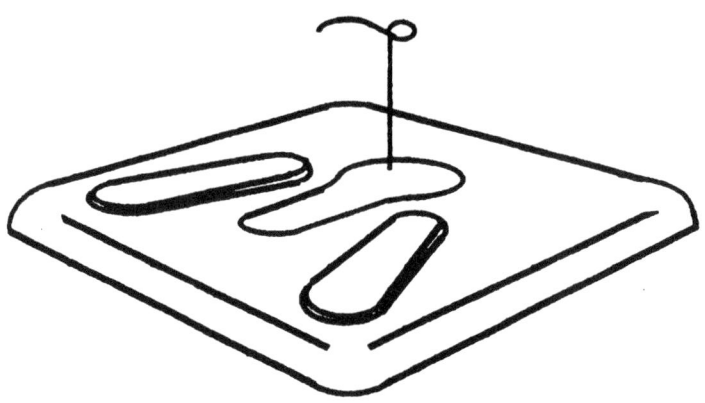

Appendix 3. Forms for sanitation programme management

In this appendix are a few forms which have been found useful in the management of sanitation programmes. The forms should be adapted to local needs and conditions as required.

Given that transport and maintenance of vehicles can be one of the key limiting factors, a vehicle inspection report form has been included.

Monthly Production and Sales Report

NAME OF PROJECT:

Council	Area	SanCentre

Production period		Form filled by	
Year	Month	Name	Date

This monthly report shall be filled in (two copies) by the managers of the SanCentres, the Council co-ordinator as well as by the Provincial co-ordinator. The second copy shall be sent to the National co-ordinator. Failure to do so will be reported as zero.

Product	Production this mth	Sale this mth	Stock	Comments

Principal activities

_____
_____
_____

Principal problems

_____
_____
_____

Proposed solutions

_____
_____
_____
_____
_____
_____

NAME OF PROJECT: _____ **INVOICE NO** ....

Invoice
Client's copy

Client's Name _____

Address/House no _____

City/Town /Village _____ **Date** _____

Description	Quantity	Unit	Unit price	Amount
Total				

Terms strictly cash

Amount in writing

_____

The strength of the slab(s) has been controlled with the load of .... persons, according to the norms of the programme.

The above mentioned amount has been received

_____

Signature by certifying officer and stamp

_____

Signature by certifying officer

Date

Household Sanitation Questionnaire

Page 1 of 2

Do not write here. Reserved for supervisor

Note 1: This form is an example only and must not be used without field testing first. Also consult the person who is responsible for the data processing.

Note 2: Remember that the field staff should be used as interviewers and that they should make their own summary and assessment of the data they have collected.

District
Sub-county
Parish
Village
Name of family
House/plot number

Interviewer Name
(In clear writing)
Year Month Day
Date of interview

1. Do you have a latrine?   Yes
   IF NO THEN GO TO QUESTION 9   No

2. When was it built?   Month Year

3. How many people use this latrine regularly   Adults | Children < 5
   M | F | M | F

4. How do you dispose of the faeces of young children?
   All children use latrine
   Put faeces in latrine
   Do not dispose of them
   Don't know
   Specify _____
   ← Other (specify)

5. Do the children wash their hands after using the latrine?   Yes / No / Sometimes

6. If yes or sometimes, ask to look at the hand-washing facility
   Is it close to the latrine?   Yes / No
   Can it be used <u>now</u>?   Yes / No
   (i.e. is it broken, does it have water etc?)
   Is it suitable?   Yes / No
   (i.e. can it serve its purpose, can children use it?)

<u>Continued on page 2.</u>

Adapted from <u>Coleman Gilroy (1991)</u>: A Monitoring and Evaluation System for the RUWASA Project, University of East Anglia for Carl Bro International a/s Denmark

7. Do people in the household sometimes defecate in the bush (e.g. when out working on the crops or when children are out playing) or do they <u>always</u> use the latrine?

    Adults
    - Sometimes defecate in the bush ☐
    - Always use the latrine ☐

    Children
    - Sometimes defecate in the bush ☐
    - Always use the latrine ☐

8. Do you think that your neighbours sometimes defecate in the bush or do they <u>always</u> use their latrine?

    Adults
    - Sometimes defecate in the bush ☐
    - Always use the latrine ☐

    Children
    - Sometimes defecate in the bush ☐
    - Always use the latrine ☐

9. If the answer to Question 1 is no, then where do people from this household defecate?
    - In the bush ☐
    - In a latrine of a relative or neighbour ☐
    - Other (specify) ☐

    Specify _____

10. If the answer to Question 1 is 'traditional' latrine then has the household considered building a latrine with a Sanplat?
    - Yes ☐
    - No ☐
    - Don't know ☐

11. Why haven't they built a latrine with a Sanplat?
    - Never heard of Sanplats ☐
    - Traditional latrine adequate ☐
    - Sanplats are too expensive ☐
    - No time to build ☐
    - Do not know how to build ☐
    - No Sanplats available ☐
    - Other (specify) ☐
    - Don't know ☐

    Specify _____

Do not write here. Reserved for supervisor

Ask for permission to inspect the latrine and the hand-washing facility. Make sure they don't clear it up and replace the lid before you see it. <u>You should report the normal state</u>.
Base your reply on observations <u>and</u> questions.

Latrine Inspection Report

One page only

Note 1: This form is an example only and must not be used without field testing first. Also consult the person who is responsible for the data processing.

Note 2: Remember that the field staff should be used as interviewers and that they should make their own summary and assessment of the data they have collected.

Do not write here. Reserved for supervisor

- District
- Sub-county
- Parish
- Village
- Name of family
- House/plot number
- Interviewer name (In clear writing)
- Date of interview — Year Month Day

1. **Type of latrine**
   - Traditional
   - With SanPlat
   - With other concrete slab
   - With ventpipe

2. **When was it built?** (Month / Year)

3. **Building condition** (Yes / No)
   - Gives privacy?
   - Gives protection from rain?

4. **Latrine condition** (OK / Not OK)
   (Latrine full, hole collapsing, SanPlat broken etc.)

5. **Lid present and in place?** (Yes / No)

6. **Cleaning materials present?** (Yes / No)
   (Brush, cleaning water, ash, sand, soap etc.)

7. **Water for hand-washing present?** (Yes / No)

8. **General condition as regards** (OK / Not OK)
   - Smell
   - Flies
   - Mosquitoes
   - Fouling (presence of faeces/urine around drop-hole)

9. **Inspect the hand-washing facility (if any). Is there any evidence that it is being used regularly?** (OK / Not OK)

10. **Inspect the compound. Is there any evidence of children's faeces which have not been disposed of?** (OK / Not OK)

Adapted from <u>Coleman Gilroy (1991)</u>: A Monitoring and Evaluation System for the RUWASA Project, University of East Anglia for Carl Bro International a/s Denmark

Weekly Vehicle Inspection

☑ If OK tick the box
Ⓧ If a problem circle the box
☒ If not applicable, delete the box

Date of inspection: ____/____19____

Name of Inspector: _____

Model: _____

Reg no: _____

Km reading: _____

Fluid levels
- ☐ Battery level
- ☐ Radiator coolant level
- ☐ Engine oil level
- ☐ Windshield washer level
- ☐ Brake fluid level
- ☐ Clutch fluid level
- ☐ Fuel tank level ....
- ☐ Spare fuel can full

Electrical check
- ☐ Headlamps low beams
- ☐ Headlamps high beams

Tail lights
- ☐ Brake lights
- ☐ Tail lights
- ☐ Reversing lights
- ☐ Licence plate lights

Turn signal
- ☐ Front
- ☐ Rear

Other lights
- ☐ Interior lights
- ☐ Warning lights
- ☐ Hazard lights
- ☐ Parking lights
- ☐ Panel lights

Meter works
- ☐ Speedometer
- ☐ Battery indicator
- ☐ Temperature indicator
- ☐ Fuel indicator
- ☐ Oil indicator

Other electrics
- ☐ Windscreen wiper
- ☐ Windscreen washer
- ☐ Heater/Fan
- ☐ Horn
- ☐ Cigarette lighter
- ☐ Fuses and Fusible link
- ☐ Are there spare fuses?

Brakes
- ☐ Brake pedal (Check free play)
- ☐ Hand brake
- ☐ Brake booster

Clutch pedal
- ☐ Check pedal travel distance

Steering wheel
- ☐ Check free play
- ☐ Steering wheel tilt lever works

Tyres
- ☐ Visual inspection
- ☐ Check wheel nuts

Tyre pressures   should be
- ☐ Front   ....
- ☐ Rear   ....
- ☐ Spare   ....

Visual inspection
- ☐ Fuel filter and lines (for diesels drain water form filters)
- ☐ Air filter
- ☐ Battery and cables
- ☐ Wiring
- ☐ Engine mounts
- ☐ Body mounts
- ☐ Shock absorbers
- ☐ Exhaust system
- ☐ Outside mirrors
- ☐ Inside mirror
- ☐ Door locks working
- ☐ Door handles interior
- ☐ Door handles exterior
- ☐ Seat adjustment works
- ☐ Seat belt works
- ☐ Seat upholstery
- ☐ Windows not broken or cracked
- ☐ Body damage (scratches/dents)
- ☐ No oil leaks

Radiator and hoses
- ☐ Radiator clean and not blocked
- ☐ Hoses in good condition

Fan belt
- ☐ In good condition
- ☐ Tension OK
- ☐ Windshield wiper blades

Leaf springs or coils
- ☐ Rubber bushings of springs
- ☐ Springs OK

Vehicle documents
- ☐ Owners manual
- ☐ ....
- ☐ ....

Repair kit
- ☐ Jack
- ☐ Jack handle with extension rods
- ☐ Handle for spare wheel
- ☐ Tyre spanner
- ☐ Pliers
- ☐ Adjustable spanner
- ☐ Fat screw driver
- ☐ Philips screw driver
- ☐ 10 - 12mm spanner
- ☐ 14 - 17mm spanner
- ☐ Flashlight
- ☐ Fire extinguisher
- ☐ First aid box
- ☐ Triangle(s) no ....

Special equipment
- ☐ ....
- ☐ ....
- ☐ ....
- ☐ ....
- ☐ ....
- ☐ ....
- ☐ ....
- ☐ ....
- ☐ ....
- ☐ ....
- ☐ ....
- ☐ ....

Comments
....
....
....
....
....
....
....

Bibliography

Literature which may be of interest for further reading has been listed, together with comments indicating the relevance of each title to the book.

Blanchard, K. & Johnson, S. (1982): *The One Minute Manager*, Fontana Paperbacks, William Collins, London.
>Through one minute work goals, one minute praise and one minute reprimands, *The One Minute Manager* tries to bring you increased productivity, profits and prosperity. Initially you have to look for one minute at the people you manage and realize that they are your most important resources.

Blanchard, K. *et al* (1982): *Leadership and the One Minute Manager*, Fontana Paperbacks, William Collins, London.
>This book has been instrumental in the writing of Chapter 11, How to Work With People.

Brandberg, B. (1985): *Manual De Latrinas Melhoradas*, Instituto Nacional de Planeamento Físico, Maputo Mozambique.
>This was the first published manual describing the construction of latrines within what today is called the SanPlat system.

Brandberg, B. (1990): *The SanPlat System*, SBI Consulting International AB, Box 217, S 530 30 TUN, Sweden.
>A presentation of the SanPlat system with reference to its implementation in African developing countries.

Brandberg, B. (1991): *Planning, Construction and Operation of Public and Institutional Latrines, Sida/Hesawa*, Stockholm, Sweden.
>A field manual based on findings from the Hesawa workshop on construction and use of Public and Institutional Latrines in Mwanza, Tanzania, 15-26 November 1989.

Brandberg, B. (1991): *The Sanitation Component of Ruwasa*, End of Mission Report for Carl Bro International AS, Copenhagen, Denmark.
>Here the SanPlat system was recommended to guide the programme out of a standstill which arose when the sanitation objectives of the programme did not materialize. Introducing the SanPlat system and negotiating with the community leadership led to a boom in latrine building. It is, however, too early to evaluate the sustainability of this approach. Experiences from Malawi show, however, that the negoti-

ated approach may be a very effective way to introduce the system and to create the demand for better sanitation.

Brandberg, B. (1993): 'Sanitation Revolution Bangladesh', *Waterlines*, IT Publications, London, Vol.11 No.4.

The SanPlat system was developed in and for African conditions. Experiences from rural areas in Bangladesh indicate, however, that there is great potential for the SanPlat system in Bangladesh, being an intermediate solution between the pour-flush system and 'home-made' latrines. An interesting experience from Bangladesh is the promotional system developed by Unicef within the immunization programme and adapted for low-cost sanitation promotion.

Brandberg, B. (1982): *A Study of Insect Circulation in Ten Latrines in Maputo.* Unpublished.

Flies were trapped passing in and out of 10 normal (non-VIP) latrines. An average of 11.6 flies per day were caught. There was no significant difference between the modes of construction or the humidity in the pits. There was no evidence that there were more flies during the hot and humid season but rather the opposite. Close to 100 per cent of the flies were metallic (*Chrysomyia*), only a few (less than 1 per cent) were grey (*Musca domestica*). The low number of flies trapped corresponds well with day-to-day observations in the area.

Cairncross, S. (1992): *Sanitation and Water Supply: Practical lessons from the decade*, The World Bank.

Where demand is not strong, the first priority of a sanitation programme must be to develop it. Whether or not demand is strong the perspective of the programme's management must be that of marketing a product rather than providing a service.
At the most basic level, effective marketing requires:
- A product that is attractive enough and cheap enough for people to want to pay for it.
- A market whose characteristics are determined by market research and test marketing.
- A delivery system to make the product accessible to potential purchasers.
- Promotion to inform customers about the product and develop demand.
- Service to build customer confidence that the product will be useful for a reasonable time.

Cairncross, S. & Feachem, R.G. (1983): *Environmental Health Engineering in the Tropics: An introductory text,* John Wiley & Sons, Chichester, New York, Brisbane, Toronto.

The book contains a distillation of much knowledge and experience and reviews problems as well as solutions. The breadth of the subject does not allow the authors to go into detail.

Coleman Gilroy (1991): *A Monitoring and Evaluation System for the Ruwasa Project*, Overseas Development Group, University of East Anglia, Norwich UK for Carl Bro International a/s, Denmark.

The survey forms used in the book are adopted from this report.

Coleman Gilroy (1992): *Monitoring and Evaluation: Role and rationale*, Overseas Development Group, University of East Anglia, Norwich UK. Unpublished stencil.

> The purpose of monitoring is fast feedback so that adjustments can be made while the process is still in operation. Evaluation is to feed back findings for future projects. This theory contrasts with the traditional concept that monitoring concentrates on data collection for later use in evaluation.

van Damme, J.M.G. (1985): 'The Essential Role of Drinking Water and Sanitation in Primary Health Care'. *Tropical and Geographical Medicine* 1985:37 Supplement S26.

> Based on information from Falkenmark and Feachem, van Damme states:
>> A summation of the scores give a rough guide to the overall relative importance of preventive measures considered:
>> - water quality — 14
>> - water availability — 22
>> - *excreta disposal* [Our emphasis] — 27
>> - excreta treatment — 23
>> - personal and domestic cleanliness — 22
>> - drainage and sullage disposal — 6
>> - food hygiene — 17
>
> Against such information it is difficult to understand the donor agencies' willingness to subsidize water and not sanitation projects.

Esrey, S. A. (nd): 'Interventions for the control of Diarrhoeal Diseases among Young Children: Fly Control'. *WHO Diarrhoeal Disease Control Programme*, WHO/1991.37.

> Long-term, environmentally safe fly control is difficult to achieve and sustain. Effective methods for short-term control, which involve the use of insecticides, are unsafe for humans and other animals. *The available evidence suggests that fly control is not feasible in many settings and that, even if successfully implemented, it is not a cost-effective intervention for national diarrhoeal disease control programmes.* [Our emphasis].

Esrey, S. A. *et al.* (1994): 'Multi-Country Study to Examine Relationships Between the Health of Children and the Level of Water and Sanitation Service, Distance to Water, and Type of Water Used', *IRC Water Newsletter* Number 227, Sept. 1994. The extract below is from this paper:

> Sanitation improvements:
> - *Health effects from sanitation were much larger than from improved water supplies and effects from improved water supplies were not always established.* [Our emphasis]
> - Flush toilets provided significantly greater health benefits than pit latrines, which in turn were significantly better than no improved sanitation. [Which may be due to the fact that we are measuring the impact of combined water and sanitation for families with relatively good hygiene behaviour.]

- Water supplies via yard and house connections were usually associated with better health compared to no improved water or public supplies; public supplies provided only marginal benefits.

Distance to the drinking water supply:
- Briefer round-trip water collection time was associated with better child health, particularly nutritional status.

Improved water sources for all water:
- Use of improved water supplies for all water needs is not necessarily better for improved health than use of improved water for only drinking and cooking needs.

Feachem R. G. et al. (1980): *Health Aspects of Excreta and Sullage Management — A state of the art review*. World Bank series. Appropriate Technology for Water Supply and Sanitation.

A bible on water and sanitation-related diseases, germs and their survival under differing conditions.

On the technology side the authors are vague and illustrated solutions are complicated and too expensive for the vast majority of people in developing countries.

The chapter on groundwater pollution was obviously written before Lewis (1980) presented his book entitled *The Risk of Groundwater Pollution by On-Site Sanitation in Developing Countries*.

Gathuma, M.J. (undated): *The Feasibility of Utilizing the Bacillus Thuringiensis (Bt) as a Biological Insecticide Against Flies in Tropical Countries, Epidemiologically Important Filth Flies and Associated Fly-borne Diseases*. Pilot Study For Evaluation. University of Nairobi Faculty of veterinary medicine Department of Public Health, Pharmacology and Toxicology, P.O.Box 29053 Nairobi.

The study is based on Kenyan conditions. In urban low-income areas the genus *Crysomyia* is the most prevalent (61%) followed by *Musca domestica* (37%) and other species (2%). Both *Crysomyia* and *Musca domestica* may be responsible for disease transmission though *Musca domestica* seems to be the more dangerous one. *Crysomyia* principally develop in pit latrines while *Musca domestica* develop in garbage. In rural areas *Musca domestica* is the more prevalent but seems to have its breeding places in animals' manure. Professor Carlberg, at the University of Helsingfors, Finland, is of the opinion that *Musca domestica* does not like human faecal matter. From a health point of view it therefore seems that garbage heaps are more dangerous than pit latrines. Still *Cryosomyia* may be responsible for serious food contamination as small quantities of germs, brought from pit latrines to food by *Crysomyia*, multiply to infectious quantities.

Grafström, G. (1983): *Våra Kvävda Kompetenser, En idébok för tvärtänkare*, Förlaget Akademilitteratur. Stockholm, Sweden.

The (underestimated) competence of man is explored in this very special book filled with thoughts for people looking for new (and old) ways of thinking and having things done. Available in Swedish only.

Green, E. (1982): *A Knowledge, Attitudes and Practices Survey of Water and Sanitation in Swaziland,* Ministry of Health, Rural Water-Borne Disease Control Project, Academy for Educational Development Inc., 1414-22nd Street N.W., Washington DC 20037.

    Green discusses an interesting point in 'the inertia of tradition' (p37) stating that some people find it distasteful to defecate in a house. Throughout Mozambique people build latrines with no roofs.

International Centre of Insect Physiology and Ecology (1990): *Pilot Study for Evaluating the Feasibility of Utilizing* Bacillus Thuringiensis *as a Biological Insecticide Against Flies in Tropical Countries.* ICIPE, P.O.Box 30772, Nairobi Kenya and Provivo Oy, Tekniikantie 17, 02100 Espoo, Finland.

    Bacillus thuringiensis, the most frequently used microbial insecticide in the world, is toxic to fly larvae, has a natural growth in manure and faeces, multiplies in latrines but is killed by sunlight, and is harmless to other animals and plants. Too good to be true?

Lewis, J. et al. (1980): *The Risk of Groundwater Pollution by On-Site Sanitation in Developing Countries,* IRWC - Report no. 01/82 c/o EAWAG, Überlandsstrasse, CH-8600 Dübendorf, Switzerland.

    An excellent publication on a very complicated subject. The problem is to draw conclusions. Though much light is shed on what are the optimal distances between latrines and water-points, the question remains unanswered. Longer safety distances have a negative impact on the quantity of water used, which we know is detrimental to health. The book discusses considering distances down to 10m between wells and pit latrines, though certain soils could allow even less.

Lorenz, K. (1974): *Aggression, Det Så Kallade Onda.* Original title: *Das Sogennante Böse.* Bokförlaget Pan/Nordsteds, Stockholm, Sweden.

    By studying the behaviour of fish, Konrad Lorenz comes to interesting conclusions concerning aggression and its importance for inter-individual relations, not only for fish but also for people. He postulates that we need to get angry and react aggressively. It is part of our nature and can be done in a constructive way.

Morgan, P. (1990): *Rural Water Supplies And Sanitation.* Blair Research Laboratory, Ministry of Health, Harare, Zimbabwe.

    In the section Rural Sanitation, the Blair latrine (VIP) and how it works is described and discussed, with general guidelines and building manuals.

    The tank and soakaway version (Septic-tank based VIP latrine) is of special interest for peri-urban and institutional use.

Morgan, P. (1992): *Report on a Visit to the Pilot Rural Sanitation Projects in Mozambique.*

    The writer praises traditional improved latrines in Mozambique, where the technology is based on small improvements on traditional solutions already in use in the area. The pilot project is a good example of how costs can be kept very low without jeopardizing hygiene and health. A small round slab with foot-rests and a tight fitting lid, 70cm diameter, or a wooden box with a lid are found to be useful improvements on the traditional latrine cover.

Nordberg, E. & Winblad, U. (1992): *Urban Environmental Health and Hygiene in Sub-Saharan Africa*. Prepared for Swedish International Development Authority (SIDA).

> The authors recommend strong support for rural development programmes combined with an increased involvement in environmental health and hygiene programmes for small and medium sized towns.

Nordberg, E. & Winblad, U. (1989): *Environmental Hygiene in SIDA-Supported Programmes in Africa: Review and Recommendations*. Prepared for Swedish International Development Authority.

> The authors state: 'Current development programmes tend to neglect the poorest half of the population' and 'one particular technology (Ventilated Improved Pit-latrine) has been advocated too widely and too uncritically'. They recommend as their first point 'that environmental hygiene be given a more prominent role and a larger share of the resources within water and health development support programmes'.
>
> The document thus implies that there should be more support for cheaper latrines.

Okun, D. A. (1987): *The Value of Water Supply and Sanitation in Development: An assessment of water related interventions*. WASH Technical report no. 14, prepared for the Office of Health, Bureau for Science and Technology, USAID Washington, DC.

> Based on figures from 1977–78, Okun rates diarrhoeas to be the major cause of disease and death in developing countries (thousands per year).
>
> | | Deaths | Diseases |
> |---|---|---|
> | Diarrhoeas | 5-10 000 | 3-5 000 000 |
> | Respiratory infections | 4-5 000 | |
> | Malnutrition | 2 000 | |
> | Malaria | 1 200 | 150 000 |
> | Tuberculosis | 1 000 | 20 000 |
>
> Figures from Table 3, Prevalence, mortality and morbidity of the major infectious diseases of Africa, Asia and Latin America, 1977–78. 'Figures do not always match those officially reported because under-reporting is great'.

Peters, T. & Waterman Jr, R. H. (1982): *In Search of Excellence. Lessons from America's best-run companies*. Warner Books, New York.

> By studying common features and differences between the 20 best-run companies in America, the authors have identified eight basic principles for successful management, useful not only for major companies but also for the manager of sanitation programmes.

Pickford, J. (1995): *Low-cost Sanitation: A review of practical experience*, Intermediate Technology Publications, London.

> This book is a must for any low-cost sanitation library. It reviews an impressive number of references combined with many years of personal experience working on low-cost sanitation.

Ruwasa (1992): *Sanitation Information*, Water Development Department, Luzira, Kampala.

Sale, C. (1930): *The Specialist*, Putnam and Company Limited, London.
> *The Specialist* is a classic, not only in latrine building but also illustrating commitment and professional pride. Read it and enjoy it!

SIAPAC (1991): *Water, Hygiene, Environmental Sanitation and the Control of Diarrhoeal Diseases in Botswana: A knowledge, attitudes and practices study*. Prepared for Ministry of Health and Ministry of Local Government and Lands.
> VIP latrines are here found to have a higher incidence of diarrhoeas than non-VIPs (Table 4.16 p43). The explanation may be found in the way VIPs were promoted in Botswana. There is however no evidence of VIPs protecting people's health better than non VIPs.

da Silva Monteizo, P. Ó. (1987): *A Latrina Séptica Escolar*, Universidade Eduardo Mondlane, Maputo.
> A study of the construction and use of the septic tank based VIP latrine for peri-urban schools in Maputo. Found to work excellently though not a long term evaluation.

Schumacher, E.F. (1973): *Small Is Beautiful: Economics as if people mattered*. Abacus, London (available from IT Publications).
> This book paved the way for the introduction of intermediate technology, a technology with a human face, as a concept (solution?) for development work not only in developing countries but for the whole world. Schumacher has the courage to mix technology and economy with religion and philosophy but, most of all, common sense.

Unicef (1991): *Unicef Eastern and Southern Africa Profile*. Unicef Eastern and Southern Africa Regional Office, P.O.Box 44145, Nairobi, Kenya.
> Contains useful health and population data for the Eastern and Southern Africa region.

UNICEF/MOH/MOW&MD/MLG (1985) *Choosing, Building and Maintaining VIP Latrines*.

van Wijk-Sijbesma, C. (1985): *Participation of Women in Water Supply and Sanitation, Roles and Realities*, International Centre for Water Supply and Sanitation, The Hague, Netherlands.
> This book discusses two sets of reasons for involving women in water and sanitation programmes.
> Economic and health benefits as:
> - The workload on the women may be reduced.
> - Health benefits for individuals and corresponding community benefits in terms of increased production and reduced costs for treatment.
> - Involving women from the beginning of the programme will in most cases lead to better maintenance and greater health benefits for the community.
>
> Project benefits through:
> - A more active community participation. Being responsible for hygiene and health in the family, women possess very useful knowledge of what the community needs and how it can be achieved.
>
> The active involvement of women may, however, be hampered by traditional concepts about what women should and should not do.

Winblad, U. & Kilama, W. (1985): *Sanitation Without Water*, Macmillan Publishers Limited, London

>An overview of many low-cost sanitation technologies illustrated by Kjell Torstensson.

World Bank (1991): *World Development Report 1991: The challenge of development.*

>The report contains useful figures on country development features.

World Bank. *TAG Technical Notes*

>For example: Nostrand, J. and Wilson, G. (1983). TAG Technical Note 3: 'The Ventilated Improved Double Pit Latrine: A construction manual for Bostwana'.
>
>The World Bank has produced a series of papers on low-cost sanitation focusing principally on VIP and pour-flush latrines. The publications have played a very important role in showing to the world that the major institutions have confidence in low-cost solutions.

World Health Organization (1983): *Minimum Evaluation Procedure*, WHO, Geneva

>This book describes a system for evaluation of water and sanitation projects without going into complicated, expensive and uncertain evaluation of health impacts. Instead the report suggests that the function and use of the physical facilities is monitored and evaluated assuming that function and good use will eventually contribute to health in the area.

www.ingramcontent.com/pod-product-compliance
Ingram Content Group UK Ltd.
Pitfield, Milton Keynes, MK11 3LW, UK
UKHW060343150426
5217IPUK00029B/2093

9 781853 393068